D0329226

SET-UP TIME REDUCTION

SET-UP TIME REDUCTION

JERRY W. CLAUNCH

Professional Publishing®

Chicago • London • Singapore

© Richard D. Irwin, a Times Mirror Higher Education Group, Inc. company, 1996

All rights reserved. No part of this publication may be reproduced, stored in a retrieval system, or transmitted, in any form or by any means, electronic, mechanical, photocopying, recording, or otherwise, without the prior written permission of the publisher.

This publication is designed to provide accurate and authoritative information in regard to the subject matter covered. It is sold with the understanding that neither the author nor the publisher is engaged in rendering legal, accounting, or other professional service. If legal advice or other expert assistance is required, the services of a competent professional person should be sought.

From a Declaration of Principles jointly adopted by a Committee of the American Bar Association and a Committee of Publishers.

Times Mirror
Higher Education Group

Library of Congress Cataloging-in-Publication Data

Claunch, Jerry, 1947–
How to increase profits and decrease costs / Jerry W. Claunch.
p.   cm.
Includes index.
ISBN 0-7863-0863-X
1. Costs, Industrial.   2. Cost Control.   3. Assembly-line methods.
4. Production management.   I. Title.
TS167.C54 1996
658.5'15—dc20   96–3988

Printed in the United States of America
1 2 3 4 5 6 7 8 9 0 QF 3 2 1 0 9 8 7 6

CONTENTS

Chapter 6

Application Improvements for Equipment   81

Chapter 7

Supporting Set-Up Time Reduction   101

Chapter 8

Holding the Gains   121

Chapter 9

What Should You Expect from Set-Up Time Reduction?   135

Chapter 10

The Team Experience   155

Chapter 11

How to Make It Fail   169

Chapter 12

Most Frequently Asked Questions   183

No longer must we simply accept set-up time as a necessary part of production. Senior managers, middle managers, supervisors, engineers, operators, and set-up employees in the past have been lulled into this truism. Utilizing the method presented in this book, you can realize a substantial reduction in set-up time which then becomes productive time and is manifest in increased profits. Too often books today tell you what you need to do without providing a method, much less one that is proven to work. The aim of this book is to provide a proven method of reducing the set-up time that you can apply to manufacturing the products of your company. This book provides all the information you will need to achieve set-up time reduction at your company. With the right focus, reduction of set-up time will provide results year after year. Through this book you will understand that reducing set-up time is only the first step; you must also analyze the benefits and make changes to the production process as well. This will allow you to achieve the highest return on investment. This process of set-up time reduction has been refined over the last 19 years and the result is this book, a proven method that will work for you.

Chapter 1 provides a list of areas that need to be addressed early in most companies. By implementing the recommended improvements, you can realize a 30 percent reduction in set-up time.

Chapter 2 will be best utilized by your financial department to assess the return on investment that can be realized as a result of set-up time reduction. Increased profit makes set-up time reduction a worthwhile initiative.

Chapter 3 presents the structure needed to support set-up time reduction. It provides a problem-solving method, and information on how to get employees involved in this initiative.

Chapter 4 provides the tools needed to achieve set-up time reduction primarily focusing on the documentation form and definitions, getting input, and videotaping.

Chapter 5 is focused on the improvement process your teams will use to implement improvements. The involvement of maintenance and quality employees is also addressed.

Chapter 6 takes the reader through specific steps of a setup and improvements you need to implement.

Chapter 7 contains information about other initiatives that support set-up time reduction and the organizational support required. Shop floor layout, departmental cooperation, and sharing of improvements are all covered in this chapter.

Chapter 8 provides methods for "holding the gains." Documentation, steering committee reviews and recognition are all key subjects of this chapter.

Chapter 9 provides methods to get improvements and overcome some of the common problems associated with set-up time reduction. This chapter presents actual application experience of the subject matter.

Chapter 10 deals with the very important aspect of getting employees involved. Both employees and management influence the success or failure of your company's set-up time reduction initiative.

Chapter 11 provides information on how to overcome failures experienced by others. In order to avoid failure, it may help to examine other companies' experiences.

Chapter 12 provides answers to the most frequently asked questions about set-up time reduction. By knowing the questions, you can be prepared with the answers.

Companies just like yours have achieved reductions in set-up time of over 90 percent, following the method presented in this book. Imagine your company with set-up times that are 90 percent lower than current times and the benefit to the organization. This book will set you on your way to achieving this important initiative.

Jerry W. Claunch

CHAPTER 1

Lessons Learned

Competitive athletes have always sought an advantage over the competition. Athletes have goals that sometimes seem impossible, but they choose to ignore the impossibility and remain focused on their vision: To be the best in the world. They study themselves, examine their eating habits, experiment with their workouts and are constantly seeking the advice of experts. In many cases, they look outside of their own arena to seek advice from professionals who can provide insight into their field of expertise. If those athletes excel and achieve greatness, they become world class. And while becoming world class was their goal, they didn't get there simply due to natural talent or willpower. The chosen few excelled due to hard work, focus, and dedication.

Similarly, companies today are striving to become world class. *World class* is defined as "able to compete with the best and win." If companies could find a way to out-distance the competition, would they pursue it? Of course they would. Set-up time reduction is the forerunner of many improvements that will put your company in the world class rank.

1

Companies need to reduce set-up time today to achieve shorter lead times, lower inventories, on-time deliveries, and (most important) the ability to change quickly.

Most information we read today presents the need for productivity increases. In most cases, we fail to achieve those increases. It's possible that because we use the phrase *productivity improvement* so often, we've become complacent about actually increasing productivity. Some people even believe we have reached the limits of productivity increases. In fact, we are just beginning. Technology is changing and so is the approach of lean manufacturing companies. Marshaling resources, both human and financial, will lead to greater results than ever imagined. It is equally important that these resources be applied in the proper areas.

Any organization can take the principles presented in this book and make significant improvements in a short period of time. Set-up time reduction is the right area for your company to spend a great deal of effort. With the experience companies have gained working on set-up time reduction, some similar steps of implementation have emerged. These steps are examined in this first chapter, but it is important to realize you cannot stop there: Your organization must recognize the need for change that never stops. Expect and encourage change, make it a way of life, and great improvements will happen.

SPEED AND RESULTS

The rate at which change occurs is as important as change itself. You cannot adopt a wait-and-see attitude to determine if the organization achieves. The phrase *pressure for results* signals the right approach but should not be confused with overpowering, belittling, or otherwise second-guessing the efforts of others. Pressure for results means we put pressure on ourselves first, then we ask the rest of the organization to exert effort toward the goal of set-up time reduction. If we succeed, the organization succeeds. If we fail, the organization fails.

The initial steps in set-up time reduction should not take very long to implement. When they do, it is because most people see this initiative as an additional task to be completed *as time permits*. If this is the case in your organization, incorporate these new steps into employee job descriptions or at least have employees report daily what they've accomplished in reducing set-up time.

There is no need to make this initiative the top priority while letting other work fall behind. Shipments must be made and problems must be dealt with daily, but if no time is allotted to the reduction of set-up time, then the benefit will never be realized, and the competition will pass you by. You should emphasize that set-up time reduction and production are equally important. Some progress can be made toward this goal on a regular basis.

Throughout the implementation, you should stress the need for every team member to spend one minute each week implementing change for every minute they spend in a team meeting. Typically the team meetings last one hour. You should also stress that this is a *minimum* amount of time. The more time spent implementing change, the faster results will be realized.

DISCRETIONARY TIME

Discretionary time can be defined as any extra time that employees have after they have completed their workload. Every employee has discretionary time, and some of this time could be used to help implement change. Encourage the use of this time for set-up time reduction. Later, we will examine the benefits of teams and the fact that no change should be implemented without first seeking team approval. Certainly, the use of available time beyond one hour can be done with the team's concurrence.

VISION

Your vision of the future in the organization will greatly impact how and when change occurs. You should start reducing set-up time immediately and ensure that progress never ends. Focusing on the vision that leads to the result you desire is key. The following vision will do just that:

SET-UP VISION

Walk up to machine
Set change parts in place (exactly)
Secure change parts (quickly)
Install tools (precisely)
Make first part (100% quality)

Whatever obstacles keep your organization from fulfilling its vision need to be improved, eliminated, or otherwise changed. Some employees react negatively to this vision because it may take hours for the current setups to be completed. The fact remains that if you simply start moving toward this vision and persist until you achieve it, then you too can become world class.

ISSUES TO ADDRESS FIRST

To assist you on the road to this vision, the following task list has been developed. If these suggestions are implemented, they will contribute to achieving the vision. This list is not the end but, rather, the beginning of a very fruitful journey. Each task is expanded upon with a set-up time reduction checklist (see page 21) to help with implementation. These tasks should be assigned by management with expected completion dates. These recommendations are applicable to almost every organization and, once implemented, will speed progress in set-up time reduction. The tasks will cover the following subjects:

Hand tools.

Fasteners.

Perishable tools.

Forming dies, tools, tool holders, and change parts.

Fixtures, dies, face plates, printing cylinders, forming tools, and chuck jaws.

Maintenance of parts, tools, and fixtures.

Lubricants, chemicals, and solvents.

Guard removal.

Scrap and rework.

Standardization of set-up procedures.

Making and delivering change parts in kits.

Establishing standard tool lengths.

Establishing standard tooling.

Establishing standard locations.

Establishing standard routing.

Where applicable, functional departments are identified to implement the recommended changes. If your company does not have the particular department that is recommended, you will need to determine which department will be responsible for the improvement and then make the assignment. In addition, the set-up employee and operator are treated as two different people in this book. In many companies, however, the operators do their own setups. If that is the case in your company, two operators should be used in the assignment.

HAND TOOLS

In every setup, hand tools are required. Far too often these tools are not available at the machine when setup begins. Also, the tools may not be organized for quick use. This is an easy problem to overcome:

1. Have the set-up employees identify the tools needed at every work center.
2. Have the set-up employees, operators, and supervisor work together to identify a central storage location at the machine.

This storage location should be clearly visible so that if a tool is missing, the employee can respond immediately to locate it or obtain an additional tool. One method that works well is to hang the tools on a rack and mark a shadow of each tool directly onto the rack where it belongs. It would not be difficult for any employee to identify the missing tool, locate it, or replace it. Any special hand tools required at a machine should be included in this system.

FASTENERS

Virtually every machine requires the use of fasteners, both in the operation and during set-up. Until the fasteners are eliminated, the amount of time spent looking for them needs to be greatly reduced. Have the operators and set-up employees identify the size, style, and quantity of fasteners required to be on hand at all times. Once they are identified, have the supervisors work with the operators and set-up employees to determine where and how to store them. A small storage cabinet at the machine may be just the solution. Fasteners are inexpensive; do not make them difficult to obtain when needed for setup. For example, one company purchased all its metric fasteners with a different color coating to make them more quickly visible from the standard inch dimensional fasteners. As you can see, the different coating colors eliminated needless time spent sorting and searching.

Later, we will discuss the need to eliminate fasteners from set-ups, but until you do eliminate them, keep plenty of them on hand and available at every machine.

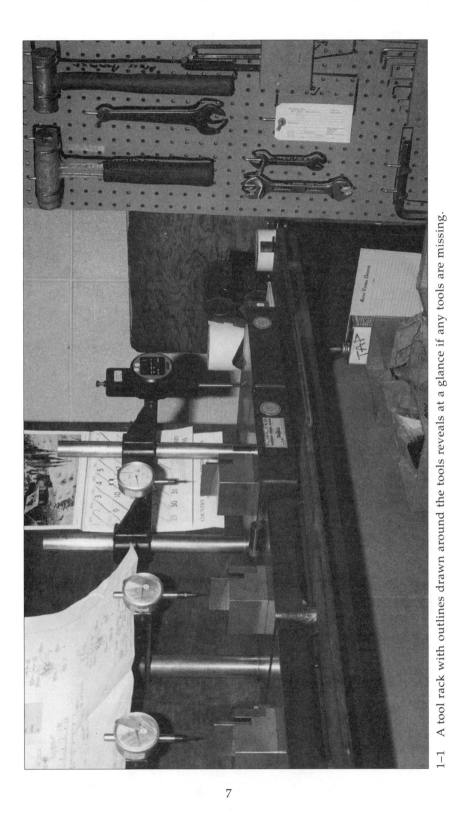

1–1   A tool rack with outlines drawn around the tools reveals at a glance if any tools are missing.

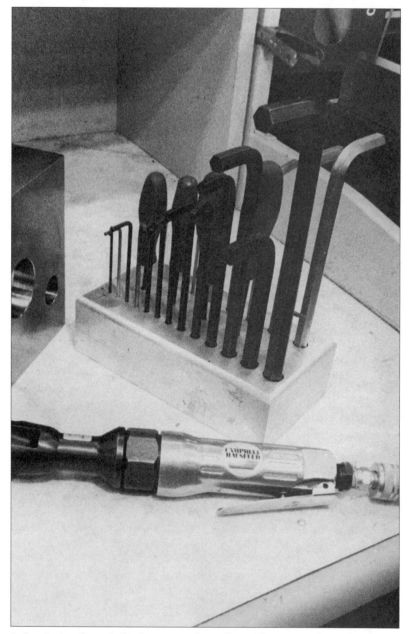

1–2   A simple rack for hex wrenches allows for a visual check to ensure that all sizes are available. As long as each slot has its wrench, this operator will have every hex wrench required for the set-up.

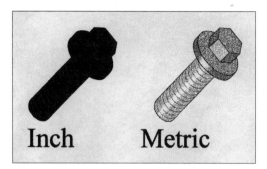

Inch   Metric

PERISHABLE TOOLS

Many machines require perishable tools to perform the job. A lathe requires cutting tools; an injection molding machine requires ejector pins; a sheeter requires cutting dies; a die-cutting machine requires cutting rules; welders require tips, and so on. Excessive time will be wasted if any of these tools are not readily available when the setup begins. Have the operators and set-up employees identify the proper quantities of all perishable tools required at each machine prior to setup. Working with their supervisor, employees should identify the most effective storage locations.

In addition, a replenishment method should be employed that will allow you to reduce the amount of perishables kept on hand at all times yet eliminate stock outs. Kanban is an inventory replenishment method that is based on standard containers replenished as frequently as they are used. Kanban of perishable tools will ensure they are available but never become overstocked. Kanban applies to fasteners as well.

FORMING DIES, TOOLS, TOOL HOLDERS, AND CHANGE PARTS

Press brakes require forming dies; metal-cutting machines require holding fixtures; automatic punching machines require tool holders; and processing lines (such as food processing, chemical processing, and bottling lines) require change parts. These items are normally changed or adjusted during setup.

The tools, holders, and parts unique to the machine should be stored at the machine in a good state of repair and ready to run. Have the set-up employees identify these parts and the quantity of each required at each machine. Next, have the supervisors, operators, and set-up employees identify the best location for storing these parts. One possible solution is to use the method described earlier for storing hand tools.

This is also a good time to have the manufacturing engineering employees standardize the fasteners. Their goal should be to have only significant different lengths. Fasteners that are ¼" different in length should never be allowed. All similar change parts should use the same size fasteners and head configuration. This will eliminate the trial-and-error method used during most changeovers to find the correct hand tool needed to change components.

FIXTURES, DIES, FACE PLATES, PRINTING CYLINDERS, FORMING TOOLS, AND CHUCK JAWS

These items may be used by various machines throughout the department or organization. In many instances, these parts are located in a central location accessible by everyone. This is, at best, an inconvenience during a setup or changeover. There are three steps to follow in this situation.

First, have storeroom employees clearly identify each of these items and label them so they can easily be seen from a distance. It is important to stress that any labeling method employed should be able to withstand normal machine process wear. Large stamps, pressure sensitive labels, etching, and metal tags are possibilities.

Second, determine where the item is used 80 percent of the time and have the supervisor, operators, and set-up employees from that area determine the best location near the machine to store these parts.

1–3 A central location for face plates reduces the time needed to locate the appropriate face plate for mounting other change parts.

The *third* step is to organize that storage area and physically move the parts into it. The storage location should clearly identify each item and its proper placement. A large, easy-to-read number in the storage location that matches the item number of the part is desirable.

Again, this system makes it easy to identify missing or misplaced items. Items in the storage location must be ready to run, not in need of maintenance. While on this topic, let's examine the subject of maintenance in more detail.

MAINTENANCE OF PARTS, TOOLS, AND FIXTURES

All too often set-ups cannot be completed due to maintenance problems. Establish and enforce a company maintenance policy to ensure that items are handled in a responsive, orderly fashion. If all departments respond accordingly, down time and delays will be eliminated. Normally this consists of filling out a maintenance work order and sending the work order, along with the item, to the maintenance department. If your company does not have a procedure for notifying maintenance of repair needs, one must be developed. Have the supervisor, operators, and maintenance employees develop a mistake-proof system for getting items repaired when needed. Once a system is developed and everyone has been trained, supervisors must intervene if problems arise. The process should be very simple and need only involve four steps:

1. Operator recognizes a need for maintenance and fills out the notification form;
2. The form and part(s) are delivered to maintenance;
3. Maintenance repairs or coordinates the effort to get the items repaired by an outside source;
4. Items are returned to their normal storage location.

1–4   These grinding wheels are centrally located, stacked on top of one another, and not easily accessible.

1–5   Now the appropriate grinding wheels to this particular work center have been relocated to give easy access for set-up.

Flow Chart of Maintenance Steps

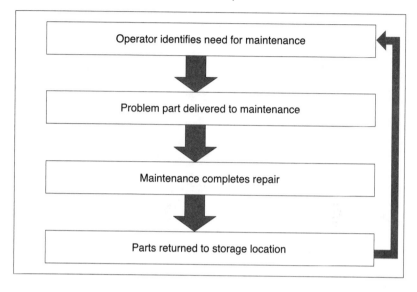

Steps must be taken to eliminate a queue in the maintenance area. The set-up time reduction concepts can be applied to maintenance as well. Items in need of repair must be fixed quickly. Otherwise you will find that missing change parts during changeover are located in the maintenance queue.

LUBRICANTS, CHEMICALS, AND SOLVENTS

Have the set-up employees and operators identify the requirements of each machine for any of these items and the quantity that should be stored at the machine. Have the safety department employees identify the proper container for the product and quantity required, as well as material safety data records. Ensure that the products are properly labeled, including any relevant hazard warnings. Once that is done, have the supervisor work with the operators and set-up employees to determine the best storage location. As shown on page 16, many companies place a rack containing the lubricants, chemicals, and solvents close to the machine.

Store Chemicals, Lubricants and Solvent at the Machine

GUARD REMOVAL

Most setups require the removal of guards, covers, or screens. These guards serve a useful purpose while the machine is in operation, but usually require the removal of fasteners. This delays the set-up.

Have the set-up employees identify the guards, covers, or screens that must be removed for setup and give that list

Application of One-Quarter Turn Fasteners

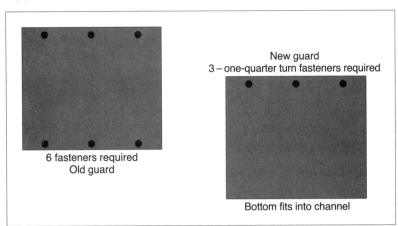

6 fasteners required
Old guard

New guard
3 – one-quarter turn fasteners required

Bottom fits into channel

to the maintenance and facilities engineering departments to develop alternate methods of attachment. The most common changes include the use of channels for the guard to fit into and one quarter (¼) turn fasteners to attach. As long as the guard remains in place during operation, the method of installation is adequate.

One quarter turn fasteners have been utilized by the aircraft industry for years with great success for attaching sheet metal to aircraft.

SCRAP AND REWORK

Too often employees accept the fact that setup requires a part to be scrapped or reworked. This situation is totally unacceptable and must not be tolerated. Have the manufacturing engineers work with the set-up employees to determine and eliminate the cause of scrap and rework. These employees may need to involve the storeroom employees or purchasing department employees to assist in correcting the problem. Normally though, eliminating such waste involves correcting programs in computer numerically controlled equipment (CNC) that provide repeatable locating and positioning thus eliminating the need for entering offsets and standardization of the setup.

Scrap or rework cannot be tolerated or overlooked. The attitude needs to be "we will find a way to eliminate scrap or rework during setup." Process documentation that clearly identifies the steps to be taken at every operation will make an improvement here. An evaluation of the gauging system and the tooling used may also help eliminate the scrap or rework. Many companies use parts that are scrapped at a previous operation as set-up parts. You should eliminate the cause of scrap at the earlier operation and discontinue adding value to a scrapped part.

STANDARDIZATION OF SET-UP PROCEDURES

As your organization progresses, you will need to ensure that all employees involved are performing the set-ups consistently. In most companies each set-up person performs the duties in his or her own way, then the operators make adjustments to get the job to run correctly. This results in wasted time.

Have the operators, manufacturing engineers and supervisors develop a standard method to be followed by all employees during setups to ensure consistency and reduce duplication of effort. A step-by-step sequential listing that can be checked off is best. Once the form is developed, have the employees document the set-up procedures on all current jobs being performed. All shifts need to be included in this effort and a consensus must be reached before implementation. This also makes it possible for one person to start the setup and another to complete it. If a setup is not completed by the same person who started it and a procedure is not followed, the second set-up person will waste a lot of time determining what tasks the previous person had completed.

MAKING AND DELIVERING THE CHANGE PARTS IN KITS

Have the toolroom employees develop a procedure to have all the change parts needed for a setup delivered to the machine in a kit. The kit should include all items necessary for the setup including paperwork. The parts that are dedicated to one job will stay in the kit. Any nondedicated parts are placed in the kit prior to delivery. After use, the kit is returned to the storage area, the nondedicated items will then be returned to their proper storage location and the remainder of the kit placed in its storage station. Ask the set-up employees to review the recommended method for kits. This allows for improvement and ownership on their part. Kits must be serviceable by the storeroom employees and meet the needs of the set-up employees.

One simple word of caution: the change parts need to be properly organized and packaged to prevent damage. If there are items that could be easily damaged, a cart with slots or cut outs for the different components is appropriate.

ESTABLISHING STANDARD TOOL LENGTHS

In machining companies, standardization of tool lengths has proven to reduce set-up time as well as simplify the programming of computer numerically controlled (CNC) equipment. The manufacturing engineering department should identify the standard tool lengths that will meet the needs of the operation. Once these standards have been identified, a system needs to be developed to ensure that tools are properly set prior to delivery.

ESTABLISHING STANDARD TOOLING

Many companies have been able to standardize the tooling used in machines, thereby eliminating the need to change tools during the setup. This includes forming tools, cylinder diameters, cutting tools, electrodes, plate sizes, chuck jaws, and so on. Standardizing tool shapes and sizes will reduce the time needed to install the tooling. This principle applies to many change parts.

Manufacturing engineering should identify any tooling that could be standardized and then proceed to implement those standard tools throughout the facility. Once the standards have been established, ensure that every setup is utilizing the standard tooling and not specials.

ESTABLISHING STANDARD LOCATIONS

If your equipment utilizes standard tooling, then standard-izing the location of those tools in the machine will further reduce set-up time. If the automatic punch machine tools are in standard locations, programming also becomes easier. The same goes for machining, notching, and forming. Standard turret positions and tool magazine locations should be identified for the tools. In the printing industry for example standardizing the colors and their location may greatly reduce the make ready time. Have the manufacturing engineers work with the operators and supervisors to determine the best locations for standardization.

Our goal in all the standardization is to reduce the need for change during setup. Certainly standardization will take time and effort, but once completed, the effort will pay back every time you have to setup. Standardization also elim-inates duplication of tooling and difficulty during setups.

ESTABLISHING STANDARD ROUTING

Part of the difficulty in setting up machines is that the rout-ings don't always follow the same sequence, even for simi-lar products. Standardizing the routings may allow for a routine setup and the more repetitious a setup, the quicker it can be done. Ask the manufacturing engineering depart-ment and shop supervisor to develop a standard routing that meets the needs of your product. Although more than one standard routing may be required, you could decide to iden-tify each family of products you produce and have one standard routing for each family.

THIRTY PERCENT REDUCTION IN SET-UP TIME ACCOMPLISHED!

All the previously listed suggestions can be implemented immediately in any company. There is no reason to wait. If

Set-Up Time Reduction Initial Issues Checklist

Item to Implement	Operator	Set-up employee	Storeroom employee	Department supervisor	Manufacturing engineering	Maintenance department	Safety department	Facilities engineering
Hand tool availability								
Fastener availability								
Perishable tooling								
Change parts								
Fixtures, etc.								
Maintenance of parts								
Lubricants and chemicals								
Guard removal								
Scrap and rework								
Standardized procedures								
Kits								
Standard tool lengths								
Standard tooling								
Standard locations								
Standard routing								

you do wait, these ideas will only have to be implemented later by the set-up time reduction teams. This means every team will have to take time to study and implement the same set of improvements in each work area. That's time wasted on nuts and bolts! By making these changes immediately and companywide, you will reduce set-up time throughout the plant and you will allow for faster improvement. By accomplishing the above list, your company may experience up to a 30 percent reduction in set-up time. This is an immediate return on investment with very little cost that is certainly worth the effort. Imagine this: all your setups are reduced by 30 percent and the additional capacity is available for production!

CHAPTER 2

Calculating the Return on Investment of Set-Up Time Reduction

There are a number of approaches to analyze the return on investment (ROI) of set-up time reduction. Ideally you have taken the initiative to implement set-up time reduction before customer demand forces you into it. In that case, you need to examine the inventory benefit. You should first consider either the cost of the money invested in inventory or the total cost to carry the inventory. It is recommended that you consider the cost-to-carry inventory since it more accurately reflects the true cost. The following worksheets can be used to determine the cost-to-carry inventory at your facility.

In order to develop all the information needed to calculate the cost-to-carry inventory, the following pages define the information needed and in some cases identify where to get the information.

Once you have developed your cost-to-carry inventory, you will now be able to calculate the benefits using the reduction model worksheet. It is important to know the cost-to-carry inventory in order to understand the benefits of set-up time reduction. In addition, this cost can be applied to other worksheets to realize the benefits associated with improvements. The effect of set-up time reduction can be far reaching as you can see from the following inventory and productivity improvement models.

You should make sure that everyone involved in inventory reduction recognizes that the reduction in inventory is *not* the savings. Since inventory is an asset, the annual cost-to-carry inventory is the savings. When inventory is reduced, the invested funds move from one asset account to another. If you choose not to consider the cost-to-carry inventory and only recognize the cost of the investment in inventory, replace the cost to carry with the cost at which your organization borrows money. If you decide to do this, you will be using a much lower number than what the costs actually are.

WORKSHEET 2–1

Cost-to-Carry Inventory Worksheets

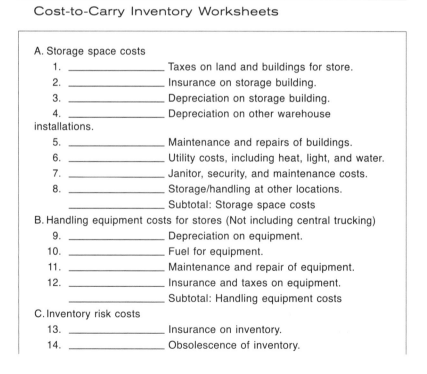

A. Storage space costs
1. _____ Taxes on land and buildings for store.
2. _____ Insurance on storage building.
3. _____ Depreciation on storage building.
4. _____ Depreciation on other warehouse installations.
5. _____ Maintenance and repairs of buildings.
6. _____ Utility costs, including heat, light, and water.
7. _____ Janitor, security, and maintenance costs.
8. _____ Storage/handling at other locations.
   _____ Subtotal: Storage space costs

B. Handling equipment costs for stores (Not including central trucking)
9. _____ Depreciation on equipment.
10. _____ Fuel for equipment.
11. _____ Maintenance and repair of equipment.
12. _____ Insurance and taxes on equipment.
   _____ Subtotal: Handling equipment costs

C. Inventory risk costs
13. _____ Insurance on inventory.
14. _____ Obsolescence of inventory.

WORKSHEET 2-1

Cost-to-Carry Inventory Worksheets *(Concluded)*

15. _____ Physical deterioration of inventory including scrap.

16. _____ Pilferage.

17. _____ Losses resulting from inventory price declines.

_____ Subtotal: Inventory risk costs

D. Inventory service costs

18. _____ Taxes on inventory.

19. _____ Labor costs of handling and maintaining inventory.

20. _____ Clerical costs of keeping records.

21. _____ Employer contribution to social security for all space, handling, and inventory service personnel.

22. _____ Unemployment compensation insurance for all space, handling, and inventory service personnel.

23. _____ Employer contributions to pension plans, group life, health, and accident insurance programs for all space, handling, and inventory service personnel.

24. _____ A proportionate share of general administration overhead, including all taxes, social security, pension, and employer contributions to insurance programs for administrative personnel.

_____ Subtotal: Inventory services cost

E. Capital costs

25. _____ Interest on money invested in inventory (Optional, see #30 below).

26. _____ Interest on money invested in inventory handling and control equipment.

27. _____ Interest on money invested in land and building to store inventory.

_____ Subtotal: Capital costs

_____ Total costs

F. Summary

28. _____ Average inventory on hand for storerooms considered in analysis above.

29. _____ % Calculate carrying cost percent by dividing average inventory (line 28) into total costs.

30. _____ % Current cost of money (%) if not entered in line 25.

31. _____ % Inven00 tory carrying costs percent (add lines 29 and 30).

WORKSHEET 2-2

Development of Inventory Carrying Costs

Use this document as a guide to obtain the information necessary for the previous cost-to-carry inventory document.

1. Estimate of share of real estate taxes paid for the part of the building and land occupied by stores facilities, including inside and outside storage.

2. Estimate of share of insurance for same areas. (NOTE: Many large firms are either self-insuring or carry a large deductible to cover catastrophes only.)

3. Annual depreciation actually claimed for the land and buildings used for stores.

4. Annual depreciation on any other remote locations or temporary locations used for storage.

5. Estimate of annual maintenance costs spent just for storage area buildings.

6. Estimate of annual utility costs spent just for storage areas. (If outside, include snow removal.)

7. Estimate share of annual expenses for security and janitorial for stores areas.

8. Annual costs incurred for storage and handling at other locations.

9. Annual depreciation on fork lifts, cranes, stackers, pickers, racks, etc. for all equipment used for handling inventory in stores areas only. (Do not include any equipment from Central Trucking, Shipping, Receiving, Operations, and so forth.)

10. Estimated annual fuel costs for above.

11. Estimated annual maintenance costs and/or maintenance contracts for items in line 9 above.

12. Annual insurance and taxes on equipment items in line 9 above.

13. Annual insurance premiums paid to cover casualty losses of stored Inventory.

14. Estimated annual write-offs due to obsolescence. (NOTE: If written off, material must be disposed of and/or destroyed.)

15. Estimated inventory losses due to scrap generated in stores handling, shelf life deterioration, etc. (Do not include shop generated scrap or material scrapped as a result of inspection or design change.)

16. Estimated inventory losses due to employee pilferage for personal use.

17. Estimated reduction of inventory value due to price decline using the lower of cost or market value.

18. Identify any taxes on inventory that have been assessed.

19. Identify labor costs to store inventory. This should include the annual wages and salaries of all individuals connected with storing inventory.

WORKSHEET 2-2

Development of Inventory Carrying Costs *(Concluded)*

20. Estimate clerical costs for data entry, cycle counting, reconciliation error correction, document handling, including time and cost of computer operation, maintenance, and reporting efforts.

21–24. Fringe benefit. (This can be combined with lines 22, 23, and figured all together in line 24.)

22. Umemployment contribution made by company.

23. Employer contribution to pension plans, life, and health and accident insurance.

24. Allocation of overhead for employees associated with inventory.

25. Generally thought of as the *cost of money.* This can be handled in line 30 as a percentage figure—but not in both places.

26. Estimate cost of interest expense on handling equipment purchases, or the interest that could have been earned on the money which is tied up in storage handling equipment. (This is related to line 9 above.)

27. Estimate interest cost (either paid—or opportunity lost) on the money tied up in storage building, and land. (This is related to line 3 above.)

SUMMARY

28. Since the cost figures developed in the above are annual and apply to the inventory investment across the year, an *average* investment-on-hand figure should be determined for the storeroom materials.

29. Develop a percentage figure by taking the total costs divided by the average inventory investment on hand.

30. Enter the current cost of money—usually thought of as the current national prime rate. Sometimes the actual cost is more or less than prime, but the prime rate may be close enough. (Do not duplicate line 25 here; it's one or the other.)

31. Total actual current inventory carrying cost percent is then the sum of line 29 and 30. (If you chose to enter the interest on line 25, then line 30 would be blank.)

For many companies today, customer demand has forced them to produce smaller lots and the inventory improvement has already been realized. The net result is that more time is spent doing setups and production time has been reduced. In this case, the following information should be used to determine return on investment.

WORKSHEET 2-3

Potential Return Analysis Worksheet

Inventory Reduction Model

Identify your annual cost of goods sold $ _____

Identify your on-hand inventory $ _____

On-hand inventory $ _____ × _____ % = _____

$\qquad\qquad$ cost to carry $\qquad$ Inventory cost

_____ × _____ % = _____

Inventory cost   Estimated % inventory   **Annual savings**

$\qquad\qquad$ reduction due to

$\qquad\qquad$ set-up time reduction

WORKSHEET 2-4

Potential Return Analysis Worksheet

Productivity improvement model

1. What is the benefit of one hour of machine time? $ _____

   (This includes labor, machine cost, and the value if the machine is producing product. Recommendation is to use total annual sales divided by number of machines.)

2. Divide the hour benefit by 60 (minutes per hour) = $ _____

   $\qquad\qquad\qquad\qquad$ (Rate per minute)

3. Identify the set-up time for the job videotaped, in minutes

   _____

4. Identify the reduction in set-up time you expect as a percent

   _____ %

5. Multiply the percent reduction _____ % × _____ = _____

   $\qquad\qquad\qquad\qquad$ (Set-up time) (Time saved)

6. Multiply the time saved in minutes _____ times the rate per minute

   $ _____ = $ _____

   $\qquad\qquad$ (Savings per set-up)

7. Multiply the savings per set-up $ _____ times the average annual number of set-ups _____ = $ _____

   $\qquad\qquad\qquad\qquad$ (Job savings)

WORKSHEET 2-4

Potential Return Analysis Worksheet *(Concluded)*

8. Multiply the job savings $ _____ times all other parts in that family _____ = $ _____

   (Annual savings)

9. Multiply annual savings $ _____ times estimated years remaining in production _____ = **Total Return $** _____

   (5 yr. max.)

The **Costs** to accomplish the above benefits are also critical and need to be considered:

1. **Fixed** costs are:

   Training $ _____

   Software purchase $ _____

   Purchases $ _____

   Consulting $ _____

   Other $ _____

   Add these up and you have the total fixed costs $ _____

2. **Variable** costs are:

   Weekly team meeting cost $ _____

   Weekly team assignment cost $ _____

   Other $ _____

   Add these up and you have the total weekly variable costs
   $ _____

   Multiplied by the number of weeks since team formed =

   Total variable costs $ _____

3. Total fixed costs $ _____ + Total variable costs $ _____ =

   Total Costs $ _____

Return on Investment:

Total return $ _____ divided by total cost $ _____ = _____ % ROI

If you could realize the benefits from one of the previous models, wouldn't you do it? If the answer is yes, you are now ready to start. If you have not completed the steps outlined in Chapter 1, you will save yourself a great deal of delay by doing so as quickly as possible. Once they are in place, the future efforts to realize set-up time reduction can move at a faster pace.

WHERE TO START

Once you have achieved the desire to get started, determining where to start is easy. Start where you have the greatest opportunity for return. If you determined that the inventory reduction model offered the most opportunity to your organization, then look first of all at the finished goods inventory. By identifying the "A" level finished goods inventory, you have found the place to start. Identify the flow of the product and the machines that handle the bulk of the product and you have the place to start set-up time reduction. If work-in-process inventory is the problem in your company rather than finished goods, then identify where the bulk of that inventory is bottlenecked and you have determined your starting place. If you determined that your company needs to increase productive capacity, then identify the machines with the lowest production and the highest set-up times. These machines are the place to begin implementing set-up time reduction.

You should recognize that you cannot make a mistake by starting set-up time reduction. For some companies, the first step is to acclimate their employees to accept change. For others it is important that the first implementation be successful. You cannot go wrong by starting. The fatal error would be *not* starting.

CHAPTER 3

How Do You Start?

The best method of achieving set-up time reduction is through teams. It is called tapping the resource within. Many of your employees have ideas that simply need to be implemented. If the ideas and solutions never get implemented, then you will continue just as you have in the past. You cannot allow that to happen. Change that makes improvement must become a way of life.

CHANGE FROM WITHIN GETS ACCEPTED

The first challenge you have is to get people to accept change. The theory that people resist change is incorrect. *People do not resist change; they resist change imposed by outsiders.* If your employees are allowed to take part in the development and implementation of change, then the change process is not only improved, it flourishes. By giving input, working to implement better methods and simply being a part of the process, people will look forward to change. For many, this is their chance to get rid of some frustration and improve their jobs.

Everyone has ideas and creativity; you simply need to funnel them into the right channels. But be careful to not make set-up time reduction a new program. Most companies today have a team improvement process or a total quality effort. You should implement set-up time reduction efforts with an action plan that complements your current improvement efforts. Help your employees see this as a continuation of a much broader scope.

TOTAL QUALITY MANAGEMENT

If you do not have an improvement process in place today, you should consider total quality management (TQM). This approach uses statistical data to make improvements by allowing meaningful initiatives to be coupled with set-up time reduction for great benefit. TQM has two pillars on which to base the future of your organization as shown in the example on page 33.

Set-up time reduction is simply a focused TQM effort. Teams should be authorized to address the time it takes to changeover. The number of teams you empower is restricted only by the number of employees you have and your need to have teams focused in other areas needing improvement. Let's begin with a definition of set-up time.

SET-UP TIME IS THE ELAPSED TIME

starting with the completion of the previous job and continuing until the new job is running at its normal efficiency rate.

Some people have proposed that this may not be the best definition. They suggest that set-up time be measured

The Basis of
Total Quality Managment

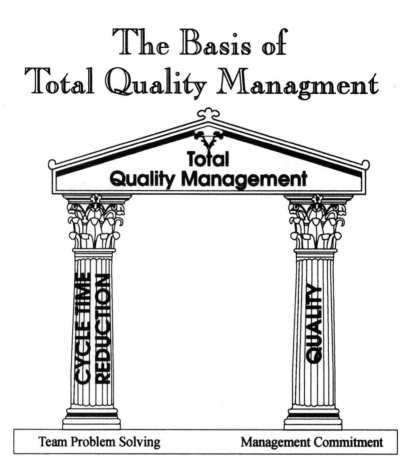

beginning with a clean machine (old fixtures and tools removed and washed down). Their concern is the amount of tooling required for the previous job. I agree with their philosophy and would have no problem if it were easy to get to a clean machine, video the setup, produce the product, and then video the steps to get back to a clean machine. While it sounds simple, most teams that try to obtain an analysis on that basis have failed to do it effectively. Later, we will discuss the documentation of a setup (which is where the approach begins to falter). Beginning with the proposed definition will enable your teams to document and improve set-up time. Using another definition may in fact make it impossible for the team to get focused and make improvements.

Don't waste a great deal of time establishing a definition of set-up time. You may find it necessary to simply request that the team agree to accept the definition and get started on their task. Even if they are adamant about another definition, they will probably change their mind and accept this definition anyway. Your goal should be to get started, and not argue about the definition.

PROBLEM-SOLVING METHOD

It is important that the teams you empower have a proven method by which they will solve problems. Too often, individuals want to implement opinions and that causes chaos. In the future, your organization will excel if fewer decisions are made by individuals based on opinions. This also includes decisions by supervision and management that are made without adequate input.

Take the time to examine the problem-solving method your employees use and determine if it meets these guidelines:

1. Reduces a problem to its possible causes.

2. Drives to root cause determination.

3. Requires more than one solution.

4. Evaluates cost and benefit of each solution.

5. Provides guidance in implementation of the solution.

If your problem-solving method achieves the above, then simply ensure that the set-up time reduction teams use that method. If your method does not provide for the above guidelines, the next few pages contain the recommended problem-solving method.

Using a proven problem-solving method will ensure that your teams get the desired result and that they don't flounder. At first, it may be difficult for the team members to follow a step-by-step approach, but once it becomes familiar, the team will follow the method without feeling confined by it (see Worksheet 3–1).

STEERING COMMITTEE

Many companies have found that it is easiest to form a steering committee to direct the implementation of set-up time reduction. If so, the organization and responsibilities of the steering committee are crucial to its success. This committee should consist of more than senior management. It should include at least two factory hourly employees, one factory supervisor, one middle manager and a senior manager. Their job is to oversee implementation, direct the teams, and provide recognition. They should not be the approval process for the team's financial needs, nor should they question decisions made by the team.

Your company presently has approval processes in place for obtaining funds. If a team needs money for implementation, it should use the existing approval process to get that funding. You may choose to allocate some amount of money to the team for purchases. A thousand dollars or less should be adequate. Beyond that, any additional funding must be approved through the established process, not by the steering committee.

In addition, if the steering committee questions the ideas and improvements the team has made, you will quickly see the team's enthusiasm wane. Eventually set-up time reduction will fade away and become one of those past improvement ideas that didn't work. So what is the job of

W O R K S H E E T 3–1

Problem-Solving Method

Step 1. Identify the symptom

Write a complete sentence identifying the symptom you want to eliminate. (This is a starting point and should be supported with a Pareto Chart.)

Step 2. Measurement of the symptom

You must now establish at least one measure whereby you can track your progress towards elimination of the symptom. You may have more than one measurement, but one is essential.

Step 3. Fact finding

List as many simple, clear answers to the following questions as you can. (Defer judgment, and use as many "why's" as necessary.)

A. What do you know or think you know about the situation?

B. What do you not know, but wish you knew about this situation?

C. What have you already tried to remedy the situation?

Step 4. Possible symptom causes

With your team, brainstorm as many possible causes of the symptom as you can think of. This should be documented in a cause and effect diagram.

Step 5. Root cause analysis

Review each possible cause one at a time and identify which other possible causes go away if that one were eliminated. Those with the most other possible causes being eliminated are root causes.

Finding the root cause is one very important step in problem solving. Having the team review each possible cause one at a time and identify which other possible causes will go away if the one they are considering were eliminated. The possible causes that eliminate the greatest number of other possible causes are the root causes. Now the team needs to develop five solutions that will eliminate each root cause.

Step 6. Solution

Here you must satisfy the following statement for every verified root cause:

Must find ways to eliminate:

_____

(verified root cause)

Solution 1:

Solution 2:

Solution 3:

Solution 4:

Solution 5:

WORKSHEET 3-1

Problem-Solving Method (Continued)

Step 7. Choose low-cost solution

Identify the root cause total cost and the total cost for each solution. (Use annual costs.)

Now compare the lowest cost solution to the root cause total cost. The solution cost should not exceed the root cause cost.

Step 8. Plan the implementation

List anticipated difficulties in getting the lowest cost solution implemented.

1. What new problems might this solution create?

2. Where could difficulties arise?

3. Who would this solution benefit?

4. How can this solution be introduced?

5. When is the best time to introduce this solution?

Step 9. Specific steps

What are the specific steps that need to be taken to verify this solution? (Be specific—the first WHO must be you.)

WHO   **WHAT**   **WHEN**

Step 10. Gaining acceptance

Develop a plan for selling the solution.

List those whom you need to convince of its value.

Name   **Dept**   **When**   **Priority**

In order to convince the above people, answer these questions:

What are the benefits of this solution?

                              Benefit   **Quantify**

How might you prove the benefit(s) of this solution?

What are the possible objections to this solution?

A.

B.

C.

D.

E.

F.

W O R K S H E E T   3—1

Problem-Solving Method *(Concluded)*

How might each objection be overcome?

A.

B.

C.

D.

E.

F.

Step 11. Refine the solution

Based on step 10, make any changes necessary to the solution. Identify the solution in full detail now:

Step 12. Implement

Now it is up to you! You need to take the actions indicated below to implement the best solution. (If you don't, nothing will happen.)

WHO   **WHAT**   **WHEN**

Step 13. Continue measurement

Measurement should continue, in order to demonstrate that the symptom goes away.

Step 14. Recognition

Determine the best way to give recognition once the solution is implemented and proven to eliminate the root cause. (At this time, the team should go to step 6 or step 1 as applicable.)

---

the steering committee? It's job is to steer the implementation, and it should take this form:

1. Grasp the vision. The steering committee should fully comprehend the need for set-up time reduction and its benefits to the entire organization. It should be able to visualize the organization after set-up times have been greatly reduced in all areas. It should focus the efforts on achieving the company vision.

2. Establish team charters. The steering committee should examine the organization and determine

where the greatest benefit would be gained in implementing set-up time reduction. It should then decide how many teams the company can support (at the start probably three to four). The steering committee then develops the team charter so the team knows exactly what it is being asked to do.

3. Staff the teams. Team staffing is crucial for the team to function without needing approval from the steering committee. Successful teams comprise all levels of the organization: senior management, middle management, supervision, as well as operators, and technicians. The team should consist of five to nine team members. The set-up time reduction team should also be represented by employees from the following areas of the plant: Set-up technicians from all shifts, operators, tool room staff, tool designers, and manufacturing engineers. Other possible areas to be involved include: quality, production scheduling, maintenance, inventory control, and stores. Start with a few teams (three or four maximum); then train them in set-up time reduction and problem solving.

4. Monthly reviews. Once a month each team should make a presentation to the steering committee, highlighting the progress it has made in the last 30 days toward achieving set-up time reduction. Again, the steering committee should not question what the team has done, unless the team is not working within their charter. The teams should present three things to the steering committee:

   - Measurement(s) of progress.
   - Documentation update.

- Problems solved.

  If the team is working on its charter and using the methods they have been trained in, the steering committee should go to step 5 and wrap up the presentation. If the team is off track, not following the methods, or otherwise not making progress, the steering committee should get them back on track diplomatically and quickly.

5. Recognition. The steering committee should give proper recognition to the accomplishments of the team. Make a big deal out of the monthly accomplishments, those accomplishments will benefit the company forever. Recognition for a job well done provides motivation to future success.

At this same steering committee meeting, after all the teams have presented their progress, it is time to have the steering committee decide whether it is time to add more teams. If the present teams are functioning well, the answer will probably be yes. If that's the case, the steering committee should revisit the vision and start the five steps again. If some of the teams are not functioning well, then it is not time to add more teams. In that case, the steering committee should work with the team leaders who need help. The goal of the steering committee is that the teams all function well, so more teams can be added.

ADDITIONAL INFORMATION ABOUT APPROVALS

In staffing of the teams as outlined above, there are senior managers, middle managers, supervisors, and any other layer you have in your organization. These people should not let the team make decisions that are not in line with the organizational goals, necessary approval processes, or any other policies and procedures. It is important that the team be staffed with layers of the organization and that the team members work to agreement. Again, the steering committee's job is to get the organization to a fulfillment of the vision. Primarily their function is to organize teams and then empower them to achieve their charter.

EMPOWERMENT

Before we look at the development of set-up time reduction teams, let's examine empowerment. Empowerment should be the goal of any improvement effort. In its simplest form, *empowered employees make team decisions, based on fact, for the good of the entire organization.* It requires that the people work together with a mission. It provides employees with greater input to factors that affect how well they can do their jobs. It necessitates removing barriers that may have existed for years. By the way, don't take lightly the "perceived barrier." If a barrier exists in a person's mind, it is real and may prevent them from moving forward. Likewise, help those who don't see a barrier to recognize that it does exist and that it may have to be eliminated in order for the team to succeed.

Empowerment goes beyond just doing a job. Team members must be willing to work with other employees to achieve higher levels of performance. In most companies, in order for empowerment to be realized, there is a need for middle managers and supervisors to change their methods as well. Dictatorial or autocratic styles are in opposition to team problem solving.

An empowered workforce recognizes that everyone has a mind, full of experiences and ideas. Once opinions and barriers are removed, those experiences and ideas can flow, through a team process, to realization. Empowerment gives more authority, once the members recognize that they also have more responsibility. Empowered teams are responsible to the rest of the organization and its employees, and because of that, they recognize the importance of the task at hand.

Empowered to Do
Set-Up Time Reduction

Your goal is to establish teams, give them a task and provide specific steps that will ensure success. Set-up time reduction teams must understand that their task is to reduce the set-up time as defined earlier in this chapter. Any efforts to reduce or improve other time elements will detract from the goal at hand. Make sure the teams don't work in other areas such as improving the production of the product, instituting preventive maintenance, and so on.

It is important to recognize that it is easy for a team to get off track. If you are not careful, teams may be working on improvements outside their charter. This may cause the organization to fail in achieving its vision.

It is also important that the team stay with the subject and not give up. Many team members are looking for "home runs." Home runs are tremendous, but few and far between. Many small improvements add up to one great one, so take what you can get, then move to the next improvement idea. Besides, small improvements are easier to implement.

TAPPING THE RESOURCE WITHIN

This is what set-up time reduction is all about—getting employees from all levels to work together, drawing on their past experiences, and implementing change that helps everyone. The goal is to let problems surface, attack and kill the problem, but not attack each other. An employee's mind is a tremendous resource waiting to be used. For some of your employees, the work they do has become mundane and provides little, if any, challenge. Set-up time reduction may be just the challenge they need. This principle does not apply to just operators. Many supervisors and managers have lost the enthusiasm and drive that got them where they are today. Everyone likes a challenge and overcoming that challenge becomes a catalyst to achieve more.

3–1   A set-up time reduction team watches and documents what they are seeing on the videotape of the current set-up method.

HOW DO YOU GET PEOPLE TO HELP?

The easiest way is to ask. Don't make light of the fact that some employees may be reluctant to participate in set-up time reduction. Henry Ford made a powerful statement about leadership: "The hourly worker will do what you ask them to do. The challenge to management is to ask them to do the right thing." Good words to manage by. This goes for senior managers, middle managers, supervisors, foremen, group leaders, union stewards, and the informal leaders in your company.

I remember doing an in-company training session on set-up time reduction and the set-up expert walked out of the classroom. He told the class he was sick and tired of young college students telling him that he had been doing the job wrong for 22 years. While I was flattered that he referred to me as a young college student at 41 years of age, I was disappointed that the person that did the set-ups was not about to be involved in set-up time reduction.

I met with him after the class and made him an offer: if he would listen to me for one hour, then I would then listen to his ideas on how to make his company more efficient. If he had a better idea, we would go to management and recommend his method be implemented and I would simply go elsewhere. Before I could begin, he said: "I don't know if I have a better idea, I just dislike being made to look bad in front of the other employees." I realized that I owed him an apology and that I should have talked with him prior to the training session. As it turned out, not only did he get involved in set-up time reduction, he became a team leader and every time I returned to the company he enthusiastically showed me what his team had done since my last visit. I have many such examples over the past 20 years where, in essence, negative people were simply turned into positive people.

TURNING NEGATIVE PEOPLE
INTO POSITIVE PEOPLE

"Will you give it a chance?" This straightforward positive question asked in a nonoffensive manner works. You don't have to like the trainer; you don't have to jump up and click your heels together at the mention of set-up time reduction. All you are asked is to spend time working as a team, helping to implement change and the result will be that your job becomes easier. At the same time you get more challenge and pride in workmanship because of the team empowerment.

I was conducting a seminar in Dearborn, Michigan, and one of the attendees, a set-up employee for an aerospace manufacturing company, told me he did not "believe any of this stuff!" I asked him to simply work with me and when the training was over to talk to me about his feelings. At the end of the session he agreed that "maybe" it would work. I challenged him to give it a try. If it didn't work after he gave it a fair chance, then I would be proven wrong. About a year later, that same employee was telling suppliers about the benefits of set-up time reduction. His comment to the suppliers was, "When this started I was totally against it, now I'm totally in favor of it. The difference is that management is right there with us working to implement change and this is the first time I have experienced this type of effort." Later that day he made an interesting comment to me. While we were talking about set-up time reduction, he pointed to a clock on the wall and said "that's my enemy . . . time. I want to do every set-up faster than the last one." He also requested that the steering committee allow him to become a facilitator for set-up time reduction.

My experience has been that if you have a proven method that works, those that are negative at first will become believers and then they will be just as much in favor of set-up time reduction as they previously were against it. The methods outlined in this book work. Just give them a chance and you will see people change.

PEOPLE ENJOY RALLYING

But people must also be consistently encouraged. In set-up time reduction, everyone should feel pressure for results. If everyone in your company feels that they had to have results in their TQM arena, then the results will happen. Top management should feel the strongest pressure for results. In order to achieve the results, you need to enlist the help of as many people as possible. Organized properly, set-up time reduction will provide excellent results.

CHAPTER 4

Getting the Results

Once the set-up time reduction teams are identified, you must provide them with the tools that will help them achieve their goal. They must be prepared to work and not see set-up time reduction as just spending time in meetings. You may hear complaints about meetings where the same things are discussed over and over, but no decisions are made. No one should tolerate meetings without results. In order to change this, or prevent it from happening in the first place, every team member must work, both in and out of the team meetings. The team meetings should be a time to review progress, discuss assignments, and make decisions. The team then makes new assignments and ends the meeting. If issues need to be resolved, then assignments should be made to gather the necessary information. Throughout a team meeting, there are two questions that every team member should ask: (1) What do we need to know? (2) What do we need to do? You will find the meetings to be effective and assignments easy to identify if the questions are asked out loud.

WHAT DO WE NEED TO KNOW?

This question is used the most, since data and information are the raw materials of decision making. Teams should not make decisions and implement change without knowing as much as they can. Your set-up time reduction teams must take the time to gather information about the current set-up methods, any problems, and opportunities for improvement. This data collection is real work and is not something done just to look busy.

WHAT DO WE NEED TO DO?

This second question drives the team to make decisions. Once all the data and information are collected, it is time to implement. Everyone on the team should take assignments. This one step, taking and doing assignments weekly, will put a team in the fast lane. A complete change in members' attitudes toward team meetings may result by simply ensuring that everyone leaves the meeting with an assignment. Let's discuss assignments in more detail.

ASSIGNMENTS

A good rule of thumb is every team member should leave the meeting with an assignment equal in length to the time spent in meetings (usually one hour per week). Emphasize that this is a *minimum* amount of time. As was discussed in Chapter 1, employees with available discretionary time should spend that time working on the team project. No one should start working on their own and decide individually what he or she will do outside the team meeting. Assignments should be agreed on and assigned to a member by the team. If a team member has additional available time, he or she should contact the leader to see if someone else needs help in completing his or her assignment(s).

There is a company in Michigan with a unique approach to getting more implementation work done. The two set-up reduction teams have projects identified that will support set-up time reduction. Each week the foreman contacts the team leaders when an employee has available time. A team member helps the employee get started on the project and follows up periodically. The assignment is still the responsibility of that team member even though it is, in essence, delegated to another employee. This has allowed the team to get more than the minimal amount of work done each week.

DECISIONS MADE

You want your teams to make decisions, and once made, ensure that they are implemented. Many teams flounder about decisions, and the entire process consumes too much time. The meeting minutes form should have a place where decisions are recorded for reference when necessary. Some teams spend a great deal of their meetings remaking decisions. A team should be careful not to move too fast, but in many cases, they are simply moving too slow. Make decisions, know the decisions you made, and ask those two questions over and over. ("What do we need to know?" "What do we need to do?") Team meetings may actually become fun!

GET INPUT FROM EVERYONE

The single mistake most frequently made is that team members do not discuss what their team is working on with other employees. Every team member should recognize the need to get input from other set-up employees, operators, and department employees before implementation. You need to limit the team size so that decisions can be made. That should not, however, prevent or limit input to the team. Team members should talk to people outside the team every week. They should discuss what the team is working on, obtain whatever information is needed, document that information, and then ask for additional input and feedback. Many good ideas are waiting to be elicited. In addition, acceptance of changes by others in the company will be facilitated if they know about the change being considered. The employees feel more owner-ship and involvement when their input is requested and used. Remember: *people resist change imposed by outsiders*. Get their input, act on it, and you are no longer an outsider.

TOOLS THAT WORK

With all the preceding information, you are now ready to begin the process of reducing set-up time. The tools are meth-ods, suggestions, and approaches that result in set-up time reduction. The intention here is to provide proven tools that your teams can use. It is important to examine the things that have worked and determine the best way for your team to proceed to get results.

BASELINING

In order to reduce set-up time, the team must know what is the current set-up. It is important that this documentation of a set-up be as accurate and complete as possible. Normally we cannot remember all the things that take place during a set-up and if you rely on memory to provide the informa-tion, you may be disappointed. There is a simple, easy way to document a set-up, which starts with making a video of a normal set-up.

VIDEOTAPE THE SETUP

A frequently asked question is, Do I have to videotape? The answer is no, but you should then ask why not? A videotape is the easiest, most accurate method to get a set-up documented. Once the tape is complete, the team can immediately document what it sees and begin working on improvement. Without a videotape, the team will need to spend more time and effort to document setup, which is a waste of valuable time.

The camera should have the ability to display time on the videotape. In many cases this is a separate attachment that may need to be added to your existing video camera. It is preferable that it display hours, minutes, and seconds. Many cameras also display the time of day.

The team should select the best job to videotape and then make the necessary preparations to ensure that an accurate video is made. No special set-up steps should be made for the videotape, but the set-up person and other operators should be notified in advance about the taping and its purpose. All efforts should be made to get a video of a normal setup, not a special demonstration. One company found that every time they filmed a setup, the time it took to do the set-up was greatly reduced from normal. The managers decided that the time it took to do a setup while being videotaped would become the new standard on the routing sheets. This deterred people from doing the special preparation that resulted in a faster setup for the video. It also provided accurate times on the routing sheets which assisted in capacity planning.

Some employees are going to resist being videotaped and you should honor their right to refuse. Usually once they see that the videotape will not be used to criticize, or harm them, they may be willing to participate in future videos. It is common in companies that have labor unions for the bargaining unit employees to be unwilling to be videotaped. In that case, the team must proceed without a video. You should never violate a contract or force someone to be in a video. Here are some guidelines that may be helpful in the use of a video:

1. Allow the set-up person in the video to retain possession of the video and be present whenever it is reviewed.
2. Never criticize the set-up person in the video.
3. Accept the video for what it is, an exact documentation of a setup.
4. Never allow anything negative on a video to be used against an employee.
5. Be sensitive to the feelings of the person in the video.
6. Obtain permission from the person in the video before it is used in training sessions.

In the event you cannot videotape a setup at your facility, the team should observe at least four setups and document what they observe during that time. This takes extra time and the team still may not have a totally accurate document, but at least it will be a good start.

When you videotape, the tape should start when the last part of the previous job is being finished. As soon as that part is complete, the timer should be started. The camera should follow the set-up expert wherever he or she goes, and should only be turned off if regularly scheduled breaks or lunches occur during the filming. If the part or parts from the project being videotaped must be inspected by someone other than the set-up expert, the camera should follow and record the entire inspection process. The camera should continue to run until the new job is running at normal efficiency.

On many occasions the first article must wait for inspection. The camera should continue to record during that time, focused on the part that is in queue. At at an aerospace manufacturing company one of the middle managers requested that we fast forward through 45 minutes of tape because the part was just waiting for inspection. The team disagreed because they felt it was useful for the manager and the rest of the team to experience the wait just like the set-up person had to during the setup. As you can imagine, the team went to work with the inspection department to figure out a way to eliminate 45 minutes from every set-up.

DOCUMENTATION OF THE SETUP

Once the videotape or manual documentation is complete, the team should then identify the elements of setup and the time they take on the documentation form. There is a copy of the form at the end of this chapter. This documentation provides the team with the ability to break the setup into elements that assist in the reduction process. The following definitions will assist in understanding the use of the form:

- Elements—All steps required in the set-up.
- Getting ready—All steps that are done prior to removing the old job, or prior to installing the new job.
- Remove old job—All steps that are involved in removing all parts, fixtures, tooling, machine changes, and clean up from the job just completed.
- Install new job—All steps that are involved in installing all parts, fixtures, tooling, and machine changes necessary for the next job.
- Prepare to run—All additional steps required prior to being able to make the first part of the next job (i.e., indicating, locating, making adjustments, or aligning).
- Trials—All steps involved in getting the first part that is accepted. If there are rejected parts made or scrap, those steps would be classified in trials.
- Buyoff—All steps in getting the first part inspected and approved.

The set-up time reduction team should identify each element of the setup first, then identify the time the element ended in the current time column. This work should be done as assignments with the team working in pairs. Each pair of team members should spend one hour minimum until the entire tape is listed in elements and current time. As the team divides into pairs for this effort, one member of the pair needs to have a good understanding of what the setup requires. The other team member writes while the set-up expert tells what is taking place on the tape. This task should take approximately 1½ minutes for each minute of tape time. The elements should be detailed enough for the team to understand what is taking place on the video.

Once the elements and current time are completed the team should, as a group, identify the activity code for each element. This can be done during a weekly meeting. The team members should not debate every element, but should be consistent in identifying activity. If they classify going to the tool room to get parts as preparation, then anytime the set-up expert goes to the tool room to get parts it, too, is classified as preparation.

Pareto Charting

The team should net the time for each element, add all the time for each activity code, and manually generate a pareto chart. With the pareto chart, the team can decide where to begin with the improvements. Since your goal is to reduce time, the activity that takes the most time is the best place to start looking for improvement possibilities.

As the team makes improvements, they should calculate the amount of improvement and put it into the appropriate category on the left of the form, along with the time. The team should keep the totals current and graph their progress weekly. With these tools, the team is now ready to make improvements to the set-up that will affect the productivity of the company for years to come.

WORKSHEET 4-1

Team name

Set-Up Video Documentation   **Date**

To External		Eliminate		Reduce		Identify Elements	Time		Classify by activity									
Yes	Time	Yes	Time	Yes	Time		Current	Net	A	B	C	D	E	F	G	H	I
□		□		□					□	□	□	□	□	□	□	□	□
□		□		□					□	□	□	□	□	□	□	□	□
□		□		□					□	□	□	□	□	□	□	□	□
□		□		□					□	□	□	□	□	□	□	□	□
□		□		□					□	□	□	□	□	□	□	□	□
□		□		□					□	□	□	□	□	□	□	□	□
□		□		□					□	□	□	□	□	□	□	□	□
□		□		□					□	□	□	□	□	□	□	□	□
□		□		□					□	□	□	□	□	□	□	□	□
□		□		□					□	□	□	□	□	□	□	□	□
□		□		□					□	□	□	□	□	□	□	□	□
□		□		□					□	□	□	□	□	□	□	□	□
□		□		□					□	□	□	□	□	□	□	□	□
□		□		□					□	□	□	□	□	□	□	□	□
□		□		□					□	□	□	□	□	□	□	□	□
□		□		□					□	□	□	□	□	□	□	□	□
TOTALS																		

ACTIVITY CLASSIFICATIONS

A. GETTING READY	B. REMOVE OLD JOB	C. INSTALL NEW JOB	D. PREPARE TO RUN	E. TRIALS	F. BUY OFF

G. VOID	H.	I.

MOVED TO EXTERNAL	ELIMINATED	REDUCED

CHAPTER 5

The Improvement Process

Once the team has completed the documentation of the videotape, they are ready to make improvements to the setup. Providing a systematic approach to this improvement will result in higher accomplishments by the team. Here is a systematic approach that your team can utilize.

The team will now consider five areas for improvement in the recommended order that they should be done, that is:

1. Eliminate quality problems.
2. Eliminate unnecessary elements.
3. Eliminate nonvalue-added elements.
4. Move elements to external.
5. Reduce all remaining internal elements.
6. Reduce all external elements.

By following these steps and working diligently to achieve them, the team will realize benefits very quickly. Let's examine each of the five areas.

ELIMINATE QUALITY PROBLEMS

The team should examine all the activities that were listed in buyoff and determine if the first part was a good part. If that first part is scrapped, or reworked, quality problems exist. Eliminating defects has a payback and simply cannot be ignored. If a quality problem exists, the team must take on the task of finding the cause and correcting it. The team can see what the set-up expert does to correct the defect and that provides insight into how to get the first part right the first time.

All you have to do is observe changeovers on packaging lines to see quality problems being dealt with continuously. The supplier of the packaging materials may be part of the cause. Having no preset positions is another culprit. Not having standardized speeds and tension can also cause quality problems. It is very common in machine shops to have set-up parts. These parts were scrapped at an earlier operation and subsequent operations use them as set-up parts. What a terrible waste. Why add value to a part that is deemed scrap? If you watched a glass forming line, you may be appalled at the lack of concern when the product had to be reground and re-melted. Printing companies historically accept waste in their process during makeready. Foundries simply re-melt the scrap and because of that, lose sight of its significance. We could go on and on, but there is no need. Every company's goal must be to eliminate defects at the source. Period!

While we're on this subject, let's elaborate. If you don't provide statistical methods in TQM, you are missing the most important ingredient. Unfortunately, there are trainers in TQM who haven't included the application of statistics. How can you make decisions or reduce variation without statistical data? If your training in TQM lacks sections on basic statistics, statistical process control and decision making with statistical data including the forms and formulas applicable to both manufacturing and nonmanufacturing areas, you should immediately make those additions to your training. It is not enough to empower teams. You must give them the tools that get results.

Many of the quality problems experienced in set-up time are due to variation. If through statistical process control (SPC) that variation is reduced, you may find that the scrap or rework during setups goes away. Providing your employees with statistical methods has tremendous payback. It is more than just the part that is rejected, you also lose all additional time which involves direct labor, machinery, support departments; and the timely delivery to the customer may be in jeopardy.

Once a team accepts its responsibility in relation to quality, it is well on its way to eliminating the cause. Team members should examine the video for clues to the cause, make assignments based on finding out "what do we need to know?" and develop solutions that eliminate the quality defects. Remember that variation from a previous operation may be the cause. Even though the team may not find the elimination of a quality problem enjoyable, it is a must.

ELIMINATE UNNECESSARY ELEMENTS

If you have taken the steps in Chapter 1 already, there may be few elements that can simply be eliminated. If those steps were not taken, your team will probably begin to address them now. The classifications of getting ready, preparing to run, and trials are key areas for finding unnecessary elements. The team can review the documentation and quickly decide if a particular element should be reviewed for elimination. As it finds elements that are candidates for elimination, it can watch them on the video and quickly determine if assignments should be made to get information and develop methods or procedural changes to eliminate that element.

It is common for unnecessary elements to be thought of as necessary by the set-up experts. The team should listen to the experts' input, and involve them in the development of new methods or procedures. Although "that's the way we've always done it" may sound funny to the team, the set-up expert has too much responsibility in ensuring a correct set-up to change quickly. Be patient and sensitive as you progress through this step. If the resistance to eliminate an unnecessary element is great, don't dwell on it. Move on. You can always come back to it as other improvements are made.

On many occasions, the question is asked: "How would you change that element?" and the feedback is excellent. Sometimes an idea emerges that wouldn't have surfaced otherwise. If a certain element can not be eliminated due to a once-in-a-while exception, the team should examine that exception for cause, eliminate the cause, then eliminate the element.

ELIMINATE THE NONVALUE-ADDED ELEMENTS

A value-added element is worthwhile doing and is completed in the least amount of time. The team should look at each element classified as getting ready, preparing to run, and trials to determine if it is value added. If the answer is no, the element should be eliminated.

Let's look at some examples for clarification. Reviewing paperwork or information is value added. Having to go get the paperwork, pulling it from a file, returning that paperwork and having to refile it is not value added. Setting adjustments in preparation to run a part is value added. Making a cut on a machined part according to the routing or work instructions is value added. Making additional cuts to get to the proper dimension is not value added. Making a setting on a printing press that you know later will need to be adjusted to the correct setting is not value added. Overfilling a pouch in food processing to make sure there are no light weights when you know that later on more adjustments will be made to get the correct weight range is not value added. In glass making, it takes time for the molds to heat up. Getting the molds to the correct operating temperature is value added, but starting with a cold mold then slowly heating the mold by running the line is not value added. Getting the glass product to the correct weight is value added, but starting with heavier than necessary weight and later on adjusting the weight to its desired weight is not value added. Here is a good checklist for nonvalue added:

- Rework is nonvalue added.
- Scrap is nonvalue added.
- Walking is nonvalue added.
- Searching for something is nonvalue added.
- Waiting is nonvalue added.
- Making repeated adjustments is nonvalue added.
- Counting is nonvalue added.
- Doing "dry runs" is nonvalue added.

The team should look for and identify all the nonvalue-added items and quickly eliminate them. There may even be many nonvalue-added elements in removing the old job or installing the new job. One example that comes to mind is when you notice a fixture that requires more than one wrench to install. Having to pick up and use different wrenches is nonvalue added. "One wrench fits all" is a good philosophy for the tool designers to live by. One fixture that a team watched being installed on video required the use of seven different wrenches. There is no doubt that the number of wrenches could have been reduced to two with very little effort. The team determined that one wrench would be sufficient.

Remember as you make improvements in your facility, be flexible, such as the number of wrenches in the above example. You might push hard to get to one wrench, but by doing so, some set-up experts or engineers may respond negatively. You may be told you can't get to one. Rather than argue or alienate that person, simply say OK, but can we reduce it to two? Being flexible and adjusting to the situation overcomes the barrier that may be developing. In many cases the set-up expert or engineer is correct, but sometimes an inability to step back and look at the situation differently causes resistance. *Be flexible but push hard for results* is a good method to employ.

As the team eliminates or reduces the nonvalue-added elements, it should update the time in the appropriate column on the left side of the form at the end of Chapter 4. We'll talk more about progress reporting later, since it is important to the success of the team and the set-up time reduction effort.

MOVE ELEMENTS TO EXTERNAL

Before using the terms *internal* and *external* let's look at the definitions of both. *Internal elements* are those elements that are done while the machine is stopped. They are internal to the machine set-up time. *External elements* are those elements that are done while the machine is running. They are external to the machine set-up time.

Whenever a team is looking to move something to external they should ask *Could* that element be done externally? Some of the team members are immediately going to question who is going to do it. Don't let your team get overly involved in who is going to do the external tasks at this point. Once the team decides an element *can* be moved to external, then they can go about deciding who should do that external element. In essence, all elements could be moved to external as long as they don't require the machine or its components to be available to complete the element. In some cases additional machine components could be purchased to move the element to external. Examples would be:

- An extra set of roll stands on a roll forming machine would allow for the rollers to be setup externally.
- An extra set of cylinders on a printing press would allow the plates to be cleaned and installed externally.
- Duplicating the filling station on a food processing line allows for cleanup and presetting to be done externally.
- Installation of a pallet system in machining would allow for the fixture to be setup externally.
- Having an additional uncoiler would allow the coil stock to be setup externally.
- Duplicating a welding station would allow for setting the weld fixture externally.
- Purchasing additional torque wrenches for assembly would allow the torque settings to be setup externally.

The team will have to justify the expense in these cases, but they may find that the expense provides a quick payback that will be experienced year after year. Most of the elements to be moved to external are simply a change in responsibility. The following is a listing of possible elements that could be done externally. The team should clearly identify when the set-up expert is doing any of the things in the following list and make every effort to move them to external.

1. Any time spent getting change parts, tools, gauges and so forth.
2. Any time spent taking tooling, fixtures, dies, and so forth.
3. Any time spent getting information or paperwork.
4. Any time spent reviewing information or paperwork.

10m x P1.5 (MOD.)
I.D. HOLDER SET SCREWS
.625 LENGTH
1.000 LENGTH
1.375 LENGTH

10m x P1.5
I.D. HOLDER & 5C SPINDLE NOSE
CAP SCREWS
20m LENGTH
25m LENGTH
30m LENGTH

WEDGES
6m x P1.0 CAP SCREWS
25m LENGTH

5C SPINDLE NOSE
SET SCREWS w/DOG POINT
5/16"-18
1/2" LENGTH
1/4" LENGTH LOCK SCREW

COOLANT LINE PARTS

COOLANT ELBOWS &
FITTINGS

10m x P1.5
S-20 SPINDLE NOSE BOLTS
.520ø SPECIAL TURNED HEAD
30m LENGTH

5-1 Here, clearly identified change parts are stored at the machine.

5–2   These computer numerically controlled tapes are centrally located and organized properly so that the one in use (see tape #14) is quickly identified. In the event storage at the machine is not practical, a central storage location like this will have to do.

5. Any time spent getting production materials.

6. Any time spent recording data (manual or computerized).

7. Any time spent cleaning up change parts, gauges, tooling and so forth.

8. Any time spent returning change parts, tools, gauges and so forth.

9. Any time spent returning information or paperwork.

10. Any time spent returning production materials.

11. Any time spent due to repair or maintenance problems.

It is especially important that the team recognize that moving an element to external is not the end. The external and remaining internal elements must still be reduced. If the team can move to external, and at the same time reduce the time to do the element, the ideal would be reached. Experience shows that these need to be done in two steps in order to get them done quickly. If your team feels they need to reduce at the same time they move to external, they may get too bogged down and not progress very quickly.

REDUCE ALL REMAINING INTERNAL ELEMENTS

The team should now look at all remaining internal elements and reduce the time they take. Every element has the possibility of being reduced, but the team should look at the elements that take the most time first. Improvements must be made in order to reduce the time. This requires creativity and brainstorming. Encouraging creativity on the part of all team members is just as important as gaining input from employees outside the team. Act creative and you will be creative should be the team's motto here.

Encourage creativity at every team meeting. Make it fun to be creative.

The team should consider ways to do each element easier, better, or faster. Every element should be examined at least once by the team. There are some specific areas that teams can examine for improvement possibilities, but do not limit the team to only these. These are suggestions and are not intended to be limiting in any way.

REDUCE ALL EXTERNAL ELEMENTS

The team should now consider ways to reduce all set-up elements that are done external (while the machine is producing). Some of these elements may have been included in the original videotape, and were moved to external by the team. Other external steps may have been done without being videotaped, and the team should consider reducing the effort on those external elements as well. Any elements can be reduced and the team should not ignore any of them, especially those that are not normally considered part of the set-up since they are accomplished externally.

PART FAMILIES

Many companies have found that they can realize a reduction in set-up time by identifying and scheduling in part families. Part families are those jobs that are very similar in set-up and if run sequentially, reduce the time necessary to set-up. Whoever plans the production schedule may be able to assist with the initial set-up time reduction efforts. By identifying and running jobs that require only minor changes to the set-up, your company may realize an immediate improvement. Many scheduling systems allow for part-family identification and then prompt the scheduler to consider the other parts in that family when a job is issued. If there is a requirement for any of those other jobs, they should be produced in sequence. This is however only a temporary improvement. The goal of your set-up time reduction team must be to reduce the time and effort necessary to do a set-up. In doing this, the need to run jobs in families may be eliminated. You need to be able to run any job, any time, in the quantity required without wasted time. Until that happens, producing in a logical sequence when you have requirements makes sense. Supervision must also understand the part families and ensure that those jobs are run in succession until it becomes unnecessary. Once the need to run in succession is eliminated, then supervision should no longer see this as a requirement.

DOCUMENTATION AND DATA ENTRY

As the team observes data entry, they should consider three possibilities: relocation of the data entry to external, reduction of the amount of data required, and finally, elimination of the data entry. Some systems require that the operator log off production then log onto setup. These two steps should be reduced to one. When the operator logs onto setup, the system should recognize that production is completed. In order to facilitate the reduction of data entry, your information system employees will need to work with the team to make this effort a reality. As the team observes manual documentation being done during the setup, they should consider if it can be moved to external, eliminated, or reduced.

CHANGE PART MAINTENANCE

Any remaining time spent finding or repairing the parts required for changeover is an opportunity for reduction or elimination. Maintenance is the first issue, since it is not uncommon for the setup to be discontinued due to lack of repair. In most companies, the maintenance staff is not allowed the time to make complete repairs. In addition, maintenance doesn't always know when a repair is necessary. The team must recognize that lack of maintenance is an improvement opportunity. A system must be developed, if it doesn't already exist, to ensure that the necessary repairs are made in a timely manner. The following suggested system does work.

At any time a change part is identified as requiring maintenance, your company notification document (maintenance work order) is filled out.

The document and paperwork are delivered to maintenance.

The part(s) are repaired or replaced and then returned to their proper storage location.

The goal of the team is that every change part is available, and ready for use.

You may have observed that the backlog in maintenance is enormous; maintenance workers are unable to get to everything especially when the backlog contains machinery and maintenance. An empowered team, working with maintenance, will find a way. Separating someone to work only on change parts is a possibility. Sending to outside suppliers to get caught up is another fix. Certainly a preventive maintenance system is a much better method. Preventive maintenance requires data collection and evaluation on a regular basis. Operators, set-up experts, storeroom employees, and maintenance employees may be involved in the data collection and analysis to ensure that preventive measures are in place. Preventive measures are well worth the effort.

CHANGE PART LOCATION/KITTING

As discussed in Chapter 1, locating the change parts at the work center where they are used is the best, but for those parts that can't be stored there, kitting is a logical method of delivering those components. The number of components

5-3   These change parts are neither organized nor identified. This means time wasted during set-up.

may be reduced, possibly through standardization. The team should consider all of these possibilities and implement in stages if necessary.

REMAINING ELEMENTS

The elements remaining are probably limited to the actual task of changing over, removal, installation, precise locating on machine, making adjustments, getting the right settings, conducting trial runs, and starting production. This can be an exciting place for the team to be because it requires creativity at its maximum. The set-up experts may have difficulty if they feel locked into the way it has always been done before. The team should look at each of the above areas in the following sequence.

Precise Locating

The act of putting the part on the machine should also locate it precisely. Further adjustments should be eliminated. In many setups, adjusting the location of a fixture, die, plate, or stop consumes a large amount of time. Many set-up videotapes cry for improvements which are available at minimal costs. There is a tremendous need to press for results in this area. The management on the team must see the need to spend a little money for such a great return. The following are some examples of changes that have been made by some teams.

Hard stops.

Dowel pins.

Ball locks.

Standardized fixture bases.

Standardized die bases.

Duplicate fixture alignment (twin spindle machine tools).

Making Settings

The goal is to make the precise setting that will be needed in order for the new job to run right. At many companies, the ability to make correct settings is impossible. It is common for temperature settings to be made using a dial ranging from one to 20. The ambient temperature affects the setting as well. You should provide a readout of the temperature at the point of use. This will enable the set-up expert to obtain the precise setting immediately and will allow the operator to monitor the setting as production is run.

Adjusting weights is another issue if required during setup. While observing the setup in a glass plant, the team noticed that the set-up expert would pick up a part and take it to a scale, then adjust a knob at the end of a long shaft. After a while, he would repeat the process of picking up a part, weighing it, and making the adjustment. Team members who asked what was going on, were told he was adjusting the weight of the part. Not having adjustment marks was the problem and the entire adjustment process was left up to trial-and-error. In no way could the set-up person accurately adjust the weights because there was no indication of how much turning the knob affected the weight of the product. A short time later, the set-up expert adjusted a damper that allowed outside air to be blown on the glass part as it was extracted from the mold. The set-up expert would watch the molten glass part and see if any distortion occurred as it was moved from the mold to the conveyor. He would observe for a while, adjust the damper, observe, adjust some more, and so on until he was satisfied that the amount of air was correctly adjusted.

These adjustments required a great deal of talent and experience. The problem is that it took too much time to get the adjustment correct. The set-up time reduction team made a few simple changes that made the task quicker. They installed a thermometer in the duct that brought the air from the outside and then recorded the amount of opening required for the size of glass part on any given outside temperature. They also added an indicator and dial on the adjustment knob and tracked the amount of weight increase or decrease when the knob was adjusted from one reading to another. Soon, the set-up people were able to make these two adjustments very quickly. If your company has speeds, feeds, or other settings that require adjustments during set-up, the team should determine if the adjustments are quick and accurate. If not, the team should find ways to accomplish this. It does not have to be computerized; it only needs to be quick and easy to adjust.

A lot of equipment has infinite adjustments. The manufacturer of the equipment makes it that way for customer versatility. Once you start using the equipment, you may find you don't need infinite settings since you may use only a few settings most of the time. Your team needs to evaluate whether fixed locations would be better than the infinite adjustments. Packaging lines, for instance, can accommodate many container sizes, but the users of the equipment may have standard package sizes and don't need the infinite adjustments.

Eliminate

There are three things that set-up time reduction teams should eliminate during the setup. The first item is fasteners. Fasteners are great for permanent mountings, but they are not desirable during setup. Fasteners require many turns to install or remove, they get lost, and they are damaged. The team should determine to find other ways to attach the change parts. In order to do that, they must understand the stress forces.

Usually fasteners are intended to resist the stresses that are imparted while the machine is producing. Far too often, fasteners receive more torque than necessary, causing damage and wasted time because of incorrectly perceived stress force. In most cases, the stress force can be overcome in other ways once the forces are understood. If there are two fasteners, possibly one could be a dowel pin and the other a fastener. Possibly a clamp that takes one quick motion can replace the fastener. If the team cannot find another way to attach or clamp, it should consider one-quarter-turn fasteners. Guards and panels should use channels and one-quarter-turn fastener.

Most set-ups require the attachment of dies, fixtures, printing plates, molds, patterns, change parts, tooling, guards, covers, air lines, power cables, hydraulic hoses, and so forth. Many times the attachment is accomplished with threads. The team should do one of the following:

1. Find low-cost methods to eliminate the threads in attachment.
2. If they can't be eliminated, reduce the number of threads.
3. If they can't be reduced, automate the method of threading.

There are many ways to eliminate threads in attachment. Quick disconnect couplers for air, hydraulic, and power lines work well provided they resist the pressure. It will be important for all your set-up time reduction teams to use the same style of quick disconnect couplers. This will reduce your need to stock replacements.

Manual clamping of parts onto dowel pins will also allow fixtures and dies to be mounted quickly and easily. Some companies have found that the installation of hydraulic clamping allows for the reduction of set-up time, while providing adequate resistance to stress forces. These hydraulic clamps can be placed on the bed of a press and once the base size is standardized, the attachment is quick. One area of caution—one company installed hydraulic clamping on their presses and about a year later while performing a continuation audit, the company found that many of the hydraulic lines were still in place, but the

5-4   This is a quick-disconnect coolant line system.

hydraulic clamps were gone, and the set-ups were being done by manual clamping with fasteners. When asked why, the team was told that the hydraulic lines started leaking and became so bad that the hydraulic clamps had to be removed. Hydraulic components, like many other items, must be maintained! Get them maintained properly and preventively. Do not allow your company to take a step backward in time.

The goal of the team is to accomplish the attachment in one motion. Any methods to do that must be safe. In order to keep the improvement safe, the team needs to determine the amount of stress force being overcome with the attachment, determine the direction of the stress force, and develop methods to overcome those stresses that can be done quickly during setup. In many cases, the team will find that the current method of attachment is overkill. The attachment would overcome more stresses than are ever present in the process. The support of the engineering department may be necessary in determining the stresses and in evaluation of proposed methods of improvement.

Dial indicators are another item needing elimination in setups. Dial indicators may be fine for ensuring quality but they are extremely wasteful in setups. Any of your setups that require dial indicators can be improved with positive positioning. One of the best ways to accomplish this task is with a product known as a "ball lock." This device allows for positive positioning of change parts with high repeatability. Relatively speaking, they are inexpensive as well. Standardizing the base of change parts enables the use of these ball locks on many setups while eliminating a great deal of time.

Earlier we dealt with the subject of infinite adjustments. The need for adjustments should be another item destined for elimination. Adjustments that are not precise and quick delay the setup. Your team(s) should find and implement changes that eliminate the necessity of adjustments.

Trial Runs

During a setup, it is common to have trial runs. The set-up expert starts the machine and tries some part of the setup to determine if the setting is correct. A printer may run a few sheets to see if the plates are lining up. In a machine shop, it is common to enter an offset and take a trial cut. On filling lines, a few containers may be filled to check the weight or level. A few parts may be run in a press brake to see how the setup is coming. Forging companies may try the first station before moving to station two. Some procedures tell the set-up expert to cycle the machine with no part or product during the trials. Some companies provide scrap parts to be used during this phase of setup. In all cases, the parts involved in the trial either become scrap, waste, or must be later reworked. Even if no part is involved, time is wasted.

If all the above were examined, the team may be able to develop methods that ensure that the first part is a good part. By doing so, the set-up time is reduced, there is no need to add value to a scrap part, and there is no need to cycle the machine without a part. The team should be empowered to eliminate the need for trials and be required to eliminate scrap, while maintaining a safe work environment during setup.

The best place to begin is to determine the location, settings, and adjustments that are in place when the job is running at production levels. If the videotape has recorded the job running at normal production efficiency, then the team can observe what had to be done to get the job to run right. If those things were in place during the set-up, the time would be greatly reduced. The team should ask what keeps us from doing that earlier? The answer identifies what needs to be overcome by the set-up time reduction team in order to reduce the trial time. If the reason is due to another department or employee not being consistent and affecting the setup, then the team can work with those people to solve the problem. Never be unsafe, but don't be overly cautious either and don't accept scrap, rework, or waste as necessary in a setup.

QUALITY ACCEPTANCE OF FIRST PART

During this phase of production, the set-up expert is usually required to check the first article part. Measurements are taken by that person to verify its acceptability to a standard and tolerance. It is common to go back and forth between trying and checking (buyoff) until the product is acceptable in the eyes of the set-up person. Once the set-up person is satisfied, the supervisor usually comes to the machine to verify the set-up person. Quality inspection usually checks the product as well. Once they approve, the set-up person may start production. Why can't the set-up expert determine if the product is correct? Probably because of some past situation wherein product was run that was not acceptable. Because of that, waste is incurred every day all year long, and you still have scrap and rework.

The application of statistical methods on the shop floor will greatly reduce the amount of time necessary for buyoff. Authority and responsibility are the key words. Once the set-up person is provided with the correct gauging (preferably variable gauging), gets the job properly centered, and charts the result, buyoff time will be reduced or virtually eliminated. Eliminating the supervisor's examination is easy. The set-up expert should make the decision as to when it is time to involve the quality inspector.

AUTOMATION

There have been times that people thought that they should be against automation or high-cost improvements. This is not the case at all, but they should guard against wasting money. If your team can reduce the set-up time with a $100.00 item why would you spend more? Many times the purchase of a technological improvement or automation will cost more than a few dollars. The term "frugal investing" may be appropriate here. Some teams start with a solution looking for a problem. We need a new computer system is an example of a solution looking for a problem. This is a very expensive way to reduce set-up time.

Automation and technology improvements should be made by the team if it is the lowest cost improvement and provides a payback in 12 months or less. There are many areas wherein set-up time can be reduced using technology. Some may be:

1. Fixture movement.
2. Clamping.
3. Automatic adjustments (DC motors, computers).
4. Pallet systems.
5. Quick change tooling.
6. Robotics.
7. New machine purchase.

If the team develops alternate solutions and finds that automation or technology improvements are necessary, they should proceed using the existing justification procedure in your company. This will ensure that the approval process is followed and that your company's current return on investment criteria are met.

STANDARDIZED PROCEDURES

As your team develops new and improved methods, and as they implement changes to the setup, it is important that all set-up employees use the improvements. It should be the requirement of the set-up time reduction team to identify the new method, develop the procedure, and arrange for the set-up experts to be trained in the procedure. It is up to supervision to ensure that all set-up employees follow the procedure, but the procedure should be simple to follow.

The improvement process can be a very exciting experience for your teams. The achievements will result in payback year after year and the set-up time reduction teams can take pride in a job well done. The steps outlined in this chapter can be followed by any team and will result in a job well done.

CHAPTER 6

Application Improvements for Equipment

If we adapt what others have done, it allows us to move quickly, but it is like the saying if you give a man a fish you feed him for a day. It will not keep you ahead of the competition. "Teach a man how to fish and you feed him for life." The goal of this book is to take your organization on the path of outdistancing the competition, and while it may be more tedious or require more work, it's what continuous improvement is all about. Without continued effort and accomplishment, you will not succeed in your ultimate goal of lean manufacturing. As you reduce set-up time, don't be surprised if your vision becomes more and more aggressive and you never stop improving. If you stop, you are as good as you will ever be, and your competitors may pass you by.

As much as possible, the information in this chapter and Chapter 7 will be related to specific industries. Challenge yourself to consider the improvement and determine how it would relate to your industry or specific application. Set-up time reduction requires you to be creative and brainstorm ideas. If you allow creativity and brainstorming, you will find this information beneficial. The purpose of this chapter is to provide some examples that you can adapt to your business.

FOCUS SET-UP TIME REDUCTION ON A PART NUMBER OR MACHINE?

If you have a particular part that you set up for frequently, focus a set-up time reduction team on that particular part. In most cases, the effort of set-up time reduction should be directed toward a machine since you want every part's set-up time reduced. If your team is focused on a machine, it should examine all of the parts that get set up on that machine.

EQUIPMENT WEAR

In food processing, the drive units that keep the carton square may need to be checked, adjusted, or replaced as part of the setup to eliminate a lot of adjustment by the operator during start-up. Teams have observed this situation on many occasions and fixing the problem properly is always much faster and economical than trying to overcome the difficulty while operating the machine. Any fasteners that typically get stripped, stretched, or otherwise damaged during the running of the product should be checked and replaced externally. This applies to all setups since fasteners are widely used to attach parts in place. In the manufacturing of glass product, the molds may become damaged or have imperfections. At the completion of the run, the operator should notify the mold repair department as to the nature of the problem and provide a sample part to further clarify the situation. The molds should then be repaired and placed in their appropriate storage area.

6-1 This milling machine has a metal guard to confine the shavings, which reduces cleanup during the changeover.

CLEANING UP

Most companies have found that the time used to clean up change over parts can easily be moved to external by getting duplicate parts that can be cleaned while the machine is producing. Before duplicate parts are recommended for purchase, the team needs to consider the cost of purchase and storage in order to determine if the purchase is justified.

Food processing examples abound: Pans used to hold product while it solidifies (such as chocolate processing) can be washed during the next production cycle if there is a duplicate set of pans, with a greatly reduced change over time. Filling units may be replaced as an entire unit as in the packaging of powdered food product into pouches (gelatin processing for example). The old unit can be wheeled into the washroom and cleaned while the line is producing. In packaging, rails that close the flaps of cartons can build up with glue. With replacement rails, the glue can be cleaned off while the next job is running.

Printing companies have reduced their make-ready time by duplicating the ink trays, cylinders, and other change parts that normally require cleaning during makeready. Some printing companies have been able to replace the entire stands including rollers, cylinders, and drive units thus enabling cleanup while the next job is running. Additionally, the amount of rags required can be greatly reduced if the parts can be cleaned in a tank with solvent. The cost of cleaning rags can add up to a lot of money. Printing companies have also found it easier to use liners for reducing the cleanup time. A liner in the ink tray greatly reduces the amount of cleanup required. Liners in containers that hold different products are beneficial as well. Cleanup is greatly reduced whether done during the setup or externally. Paint and chemical manufacturing companies have also duplicated the change parts that require cleaning in order to clean them without interrupting the next job. Pigment and other powder manufacturing companies have duplicated the change parts as well.

ADJUSTMENT

Teams constantly see adjustment knobs that lack clear identification of the amount of adjustment being given. During setup, this requires a great deal of time. As we discussed in Chapter 5, temperature adjustments for sealing equipment and packaging lines also lack indication of the exact temperature. Without the adjustment indication, the trial-and-error method consumes way too much time during setup, and contributes to lost time.

With some amount of correlation analysis, indicators can be installed that more precisely indicate the amount of adjustment being made and the expected result. You should recommend against dials that have no relationship to the adjustment such as 0 to 20. Dials that identify the amount of adjustment in standard increments should be the method of choice. In food processing, the opening of a flat carton may require less adjustment if fixed settings are used rather than infinite adjustments.

Let me make one point about overadjustment. Many times the efforts of the set-up expert and the operator amount to overadjustment of the process. The adjustments cause rework or scrap. This is then construed by others in the company as "they don't care." That is usually not the case; overadjustments are common among operators that care a great deal about the output of their process. Without statistical methods of control, the overadjustment may indeed cause quality problems later on in the manufacturing process.

FASTENERS

Fasteners should be eliminated in setup at every opportunity. They cause many problems and delays. Fasteners get lost or damaged, the hand tools get dropped, and none of this leads to quick setups. There are three simple steps to achieve fastener elimination.

- Get plenty of the fasteners currently used and keep them at the using machine.
- Get rid of all the bad ones.
- Brainstorm how to eliminate the need for fasteners altogether.

A printing company found that when plates were installed, the fasteners used on the cylinders were either damaged or missing. They also found that many of the threaded holes in the cylinders were stripped. In some cases, the pressmen were lucky to find two fasteners that would work on each cylinder. They first repaired all the threaded holes, got plenty of fasteners at each press, and then went about eliminating the fastener requirement altogether with the use of locking pins.

One Quarter Turn Fasteners

This is the only fastener that I recommend. No company should be without an adequate stock of one quarter turn fasteners on hand at all times. Once available, their usage will be great. Any time there are guards or some other cover to remove, it may greatly reduce the set-up time by incorporating one quarter turn fasteners instead of the threaded fasteners currently used to secure them. Aerospace has been using them for years to hold sheet metal onto airplanes. As long as they resist the stress forces, they function quite well.

TOOLING NEEDS

In order to determine the amount of tooling required for setups, the following steps should be followed.

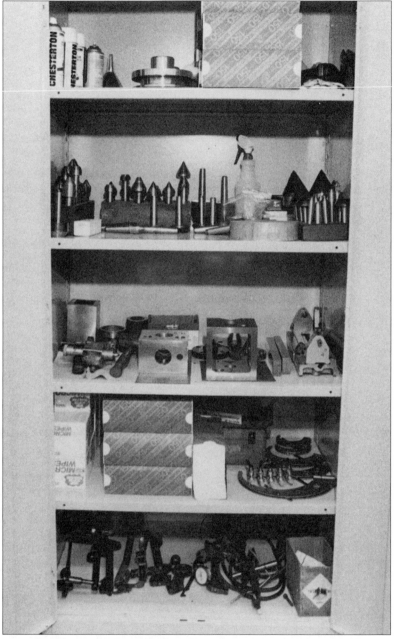

6–2 This cabinet, located behind the grinder, has selected articles that will be used during setup neatly organized.

1. Survey all operators and set-up employees to identify all required tooling.
2. Determine the storage location for the tooling. (Close to the user is the goal.)
3. Determine the required quantity of each tool. (Once the tooling is stored in its proper place, the correct quantity will be easy to determine.)

The rule to follow is that all tooling must be available when needed. Any tooling not available because of other use means additional tools need to be purchased. If you don't follow these steps, you may find that you purchase more tooling than is actually required.

Your set-up time reduction team should also identify any special tooling that is devised by either the set-up expert or the operator. This will reduce much of the trial time. This tooling should be stored close to the using work center and maintained properly.

TOOLBOX CLEAN OUT

Almost without fail, companies find the need to do a toolbox clean out early on. Many tools needed during setup are in the locked toolboxes of employees. Allowing the tools to be returned without repercussions gets them out of the limited use into the correct use, and may reduce the number of purchases required. By the way, employees don't horde tools to cause problems, they horde tools because they aren't always available when they need them. If a tool is not always available don't just purchase more. Unavailable tooling is usually attributable to where it is stored as well as the quantity on hand.

TOOL STORAGE

Some companies, as a result of set-up time reduction efforts, have found that they store tools improperly. One company had a tool with an inner cylinder that would freeze up. The team contacted the tool manufacturer and found that the cause was improper tool storage. They laid the tool on its side during storage when it should have been stored upright. One side benefit this company gained as a result of this work with the supplier was a tremendous cost reduction. The supplier came in to work on storage and further investigated the use of the tool. This company had been purchasing special tools costing $3,800 each. The supplier recommended and the company implemented their standard tooling which cost $700 each. This savings is the result of the set-up time reduction team's efforts.

QUICK DISCONNECTS

Quick disconnect chuck jaws and quick disconnect tool holders are available from many companies. Quick disconnects for airlines and vacuum lines are available as well. All quick disconnect ideas should be investigated by set-up time reduction teams for implementation. Many teams have developed their own methods of quickly disconnecting or connecting during setup.

STANDARDIZED COMPONENTS

By utilizing standardized fixtures that accommodate many of the parts manufactured, one team at a machining company greatly reduced the set-up time. They made a fixture that held all the parts in a family. These parts were gears that had to be bored and have a keyway machined. This standard fixture eliminated 14 setups. Now a setup that took 35 minutes can be done in 11 minutes. In CNC machining, standard tooling can be located in the turret or tool changer and never have to be removed or replaced during setup. As long as the tool is in a condition to run, storing it right at the machine is the best practice.

ALIGNMENT

The time spent aligning during set-up time can be reduced or eliminated in most cases. Many machine shops mount two fixtures on the table and alignment of those fixtures takes a great deal of time with a dial indicator. A simple straight edge for a stop may suffice, as was the case at one company. Dowel pins that locate a tool or fixture are also utilized by many companies to reduce the alignment time required by most setups.

In the printing industry, a pin-register system may provide the benefit to reduce or eliminate the adjustments necessary during setup. One printer found that using the same pin-locating system to locate the printing plate onto the cylinders greatly reduced the amount of adjustment for alignment. One side benefit was that they found the system that makes the plates was not calibrated correctly. Registration was always assumed to be exact, plate-to-plate. It wasn't and after corrections were made, less time was spent dealing with alignment problems during setup (makeready).

Patch-up in the die-cutting process should be considered an adjustment that needs to be eliminated. Measuring and monitoring the height of the cutting blade stock may in fact eliminate the patch-up required. Patch-up is a time-consuming nonvalue-added task during setup.

In machining it is common to make adjustments at the machine or enter offset adjustments if the equipment is computer numerically controlled (CNC). By standardizing the tooling and fixture components, in addition to presetting the tools, these adjustments can be eliminated.

Many machine shops have greatly reduced or eliminated the dry-run time wherein the machine is cycled without a part in the fixture to ensure that there will be no crashes. The operator normally does the dry runs one at a time and verifies each tool. Doing the dry runs all at one time provides some improvement. Don't forget, once the causes that necessitate a dry run are all eliminated, the dry run can be eliminated as well. Standardizing the CNC programming procedures will help reduce the dry-run time (i.e., always rapiding to .100" above a machined surface allows for a quick .100" gauge block check of location). Presetting the tooling is another key factor in eliminating adjustments.

TOOL AND FIXTURE POSITIONING

It becomes apparent as soon as you watch a videotape that in most cases the time to position is much greater than it should be. The first step is to observe if the operator changes the positioning at the beginning of the run. If so, that position should be documented and achieved during the setup.

Second, dowel pins and hard stops provide a simple and quick repetition of positioning. You must remember that dowel pins wear and need to be replaced periodically. Numerous setups can be simplified by the use of hard stops. One employee was putting two fixtures on a twin spindle Matsurra machining center and used a "Magic marker" to scribe a line to facilitate aligning the two fixtures. He then used a dial indicator to get them to within .0001". What a tremendous waste of time. A hard stop on the table would have allowed the set-up person to make the alignment very quickly with high repeatability.

Ball locks are an excellent way to establish and repeat the positioning of fixtures. Normally, in order to use them, the fixture is put onto a plate and the plate is then machined to allow the ball lock to position into the receiver which is installed into the table or bolster plate. The number of ball locks required depends on the stress forces that will be encountered. Many of the perishable tooling manufacturers offer ball locks.

On computer numerically controlled (CNC) equipment, the use of probes may reduce the set-up time. The probe is used to determine the exact location point that is used by the program in the machining process. The probe touches a known surface for location and the tools use that location as a reference. Probing should only be done during setup to avoid adding unnecessary cycle time.

Standardized tooling allows the set-up time to be reduced since the tools don't have to be replaced during setup. Some companies have been able to standardize 50 percent of the tools, while others have 100 percent standardized tooling. In either case, the set-up time has been reduced and standardized tooling has proven beneficial.

Another method to repeat and position precisely is pallet systems. Although more expensive, they are very quick and are more adaptable to situations where the fixture is used frequently. A pallet system requires a receiver be attached to the machine while the fixture is attached to the pallet. The pallet holding the fixture is attached to the receiver by air vacuum.

The use of a tombstone, on which fixtures can be permanently mounted, greatly reduces the positioning time. Normally one side is left empty in order to accommodate those fixtures that are used infrequently. A tombstone has four sides upon which frequently used fixtures are permanently mounted. Silos with cheese board is another variation of this approach. The silo is mounted to a pallet in a flexible machining center and fixtures are permanently mounted onto the silo.

CLAMPING

The time it takes to clamp a fixture, tool, or other component to the machine during setup is a prime candidate for improvement. Finding alternate methods is well worth the effort. DeStaCo-type clamps, hydraulic clamps, and air-over hydraulic clamps are some possibilities that can be installed to reduce the set-up time.

In job shops it is not always possible to standardize the clamping since fixtures and positions can vary based on the machine that is available to run the job. Most job shops find that the time the set-up person spends looking for the clamps or other components such as fasteners can be greatly reduced if kits consisting of a variety of clamps and fasteners are available at every work center. It is also recommended that the engineers standardize the base thickness of fixtures so the set-up time reduction team can standardize the clamping height and position.

TOOL PRESETTING

Presetting of tools can greatly reduce the set-up time. If you aren't already doing it, you may find it beneficial in your company. One CNC machining company had operators do their own setups and, after evaluation of the set-up videotape, determined that the setting of tools consumed the bulk of the time. They identified one operator and asked him to be the full-time presetter and the initial results show great potential. The tool setter is faster since that is his full-time job, plus the quality of work is greatly improved. Likewise the setups are much faster and the volume of work produced has increased—all this without adding a single employee.

TRAVEL TIME

It seems that every set-up time reduction team begins with a tool cart. All change parts are located in the cart along with the hand tools. Most videotapes of setups will show that there is a great deal of time spent traveling to get things needed during the setup. One food processing plant purchased some storage units on wheels and put on them all the items needed for a changeover (all tools, change parts, testing equipment, etc.). Today the set-up expert simply brings

6–3   Preset tools are ready to be moved to the machine for setup.

the storage unit to the processing line and does a changeover with a lot less time lost due to travel.

A machine shop team established runners to get items needed by the set-up employees. Initially the team put lights in each cell and when the light was turned on, a runner came to the cell and went to get whatever was needed. This allowed the set-up expert to keep working. After a few months, the team justified radios that eliminated the runner having to come to the cell to find out what was needed. The set-up expert simply radioed the need to the runner who delivered the required items. Later, kits were implemented that reduced the number of missing components needed for the changeover. The runners also assisted in bringing materials such as cutting tools and raw materials that were needed during production.

TRIAL RUNS

Trial runs are a tremendous waste of time. Why cycle a machine to make sure the tool doesn't interfere with the fixture, part, or other obstacle? By all means, you should never create an unsafe environment wherein crashes become common place, but if you eliminate the cause of the crash, then the trial run becomes unnecessary. Standard fixtures, locations, and preset tools may in fact make trial runs obsolete.

One machining company team watched their video of a setup and found that the operator would do one feature of the operation, remove the part, inspect that feature, and then make any necessary adjustments to the machine. The operator would then put the part back in the fixture and perform the next feature, remove the part, inspect, and adjust if necessary. This process continued until all features at the operation were completed, inspected, and necessary adjustments made. By simply doing all the features, removing and inspecting the part, then making all the adjustments, the team reduced the set-up time greatly.

If a job has run before, there should be no need for trials. If your company has a computer aided manufacturing system, it may have the ability to do a computer-simulated trial that would prevent the need for a trial during setup. Keep your factory safe, but eliminate every trial you can.

STRESS FORCES

Because most companies have not adequately identified the stress forces that will be encountered during the manufacturing process much set-up time is spent over clamping. By determining the stress forces, a clamping method can be specified that will be adequate, not over tightened, and take only the necessary time during setup.

On many occasions you will find over fastening, probably due to the philosophy that if a little is good, a lot is better! The problem is that the set-up expert must find all the clamps, fasteners, hand tools, and so forth and take the time to install them. If the clamping is unnecessary, the time is wasted each time the setup is done. The engineers should be able to assist in determining unnecessary clamping and what stress forces should be overcome. The team can then implement the correct clamping and eliminate what is unnecessary.

DEDICATED EQUIPMENT

Dedicating equipment to the manufacture of a high-volume part or product may be beneficial to your company. One machining company found that it incurred no setup by dedicating a manual machine to one part number that ran frequently. They were going to eliminate the machine prior to a set-up time reduction team determining and justifying the change. Another machining company found that they could reduce the set-up time by assigning magnesium parts to run on different equipment than the aluminum parts which require a different coolant. Prior to this, the coolant was changed during the setup, requiring 30 minutes. Today, changing the coolant has been eliminated.

A food manufacturing plant found that dedicating equipment to the highest volume flavors allowed them to eliminate set-up time for two flavors while the three other machines made the other 19 flavors. The set-up reduction team reduced the change over time on the other three machines, so all flavors could be run weekly.

EQUIPMENT MAINTENANCE

Many set-up time reduction teams have identified maintenance as an improvement. Far too often the maintenance and repair of equipment is taken for granted and the maintenance department can not get the job done. Teams should be on the lookout for opportunities and work to get preventive maintenance in place. This includes inspection equipment, machines, gauges, fixtures, tools, clamps, guards, and all other change parts.

Changes to equipment are a natural by-product of set-up time reduction and the examples are many. A sheeter operator determined that some of the variation in size was due to the centrifugal force of the roll stock. He determined that a dancer roller would help reduce the pull, or push, of the roll as the paper was being sheeted. The result was a higher-quality product and reduced set-up time in the press area.

Having an operator come in early and start the equipment that requires warm-up may greatly reduce the variation experienced after the machine is down for a period of time. In addition, some companies have an employee come in early to oil the equipment.

Improvements to the equipment are a necessary part of set-up time reduction. Your teams' efforts in this area will provide results that will benefit your company for years to come. It is not enough to copy what others have done; your improvement efforts must take your company beyond the competition. These ideas have come as a result of many teams working for many years to reduce set-up time. They in no way exhaust all the possibilities, and your teams will develop many improvements beyond those covered in this chapter.

AUTOMATION

Automation is not a first step. Too often, the solution that team members develop and recommend to reduce set-up time is automation, which is very expensive. Finding the low-cost solutions that reduce set-up time is the correct first step. If automation is the low-cost solution, then by all means automate. Let's look at some of the examples of automation that have been implemented by teams.

Many machining companies have purchased and implemented pallet systems. Pallet systems not only reduce the set-up time, but also allow parts to be mounted and moved to different equipment without removal from the pallet. A pallet system consists of a receiver that is mounted on the machine and the pallet on which the fixture that holds the part is mounted. In addition to making the setup very fast, the run time can be reduced by having two pallets where the operator can unload and reload the pallet while another part is being machined. The receiver is permanently mounted. A pallet would be required for each fixture used to hold the parts run on the machine.

If you have a lot of parts to run on the machine, the investment in pallets may become significant unless the pallets are setup externally. In this case, you could purchase a minimal amount of pallets that are preset external to the setup by a qualified employee. This reduces the number of pallets required and still allows you to utilize the pallet system.

You should consider investment in air-transfer systems to handle the heavier dies, fixtures, or plates that you have to change during setup. Many commercial systems for air transfer are available in the marketplace and they greatly reduce the time spent waiting for lift trucks or cranes during setup. In addition they normally make alignment of the fixture, die, or plate easier.

Any time heavy items are moved by hand, your team should consider providing a lifting device that will make the job easier or quicker. Hydraulic lift tables, roller conveyors, and ball transfers are simple improvements that may reduce strain as well as set-up time. Whatever method the team implements, safety should be a primary concern at all times.

In some cases, teams may find that the implementation of robotics may in fact reduce set-up time. This may include the use of automatic guided vehicle system for delivery of a kit or the raw material. Implementing robotics in the manufacturing process may make changeover quicker due to the preprogramming of the robot's computer to run the new job. Implementing a direct numerical control (DNC) system for loading programs into the computer numerically controlled (CNC) equipment may greatly reduce the set-up time. Usually the next job program can be downloaded externally thus reducing the set-up time.

CHAPTER 7

Supporting Set-Up Time Reduction

QUALITY IMPROVEMENTS

As you observe the videotape of your setups, you may find that much of the start up time can be reduced if specific things are done during the setup. If these are not done, usually the result is scrap or rework of the product produced during the setup.

There are some generic improvements that affect quality. They should be considered early on:

1. Consider the lighting, is it adequate?
2. Eliminate inspection queues, if possible.
3. Determine whether gauges are available when needed.
4. Determine whether the gauges are adequate to meet the quality specification requirements.
5. Conduct training in the correct use of gauges.
6. Determine whether the routings and work instructions meet the needs of the users.
7. Eliminate any supervisor approval of quality.

7–1 These gauges have been made available, preset, and ready to be used during setup.

The quality improvements in a preceding operation may also affect the set-up time. In sheet-fed printing operations, the size variation and out of square condition of the sheets can cause numerous problems during the press makeready. Reducing variation by statistical process control (SPC) at the sheeter can greatly reduce make-ready time and improve the quality during the run.

One printing company eliminated the scuffing of packaging materials as they were being folded by eliminating the chain gripper that forced the carton onto the belt. The set-up reduction team designed a tapered guide that used air pressure in place of the mechanical chain drive. They chrome plated the guide to further eliminate sticking and damage. This same company implemented lot control of the raw materials which allows them to help control color. With lot control, the team can continue to use the same lot of raw materials as it runs once the color is set during the makeready.

A printing company team was working on the length repeat of the web on foam labels and determined that during the makeready, they should plot the length using a control chart. After the team saw the variation, they brainstormed what had caused it. They determined that when they ran certain jobs, the blanket caused the problem. They now replace the blanket during every makeready and are able to control the length of the foam label. In tracking dot gain and density, a printer developed process control and today the pressman knows when *not* to make an adjustment. In most cases, maintaining quality should not require a lot of adjustments.

One company was collecting data for statistical process control and determined that there was an assignable cause for the size variation in the sheets of paper to be printed. After investigation the team determined that a gear was missing part of one tooth. After replacing the gear, set-up time was also reduced due to the continuous engagement of the new gear.

Many companies require that the supervisor be involved in quality inspection, and that takes them away from supervising. Many times in the documentation of set-up time we see three quality inspections in sequence. The operator inspects the first article, the supervisor inspects, and then someone in the quality department inspects. Your team should consider eliminating supervision from this loop, freeing them from this nonvalue-added task. If they are finding errors or quality problems that the operators are not finding, training is the answer. Your goal should be to solve the problem instead of treating the symptom with supervisor quality inspection.

One machining company found that a lot of the height stands throughout the shop were missing. The result was time during setup spent looking for stands. The team secured the funds to replace those that were either lost or damaged and got the others repaired. Some of the improvements are simple, commonsense issues but they don't get done by magic. Another machining company found that the set-up employees were spending a lot of time attaching the dial indicators to the height stands because the clamps were missing. Replacement clamps were purchased and installed by the team.

VALUE ANALYSIS/VALUE ENGINEERING

It is a shame that value analysis, also called value engineering, has lost its spotlight as a method to reduce cost without reducing quality. During the 1970s and early 1980s value analysis contributed to profit improvements at many companies and many opportunities are still available. Applying value analysis may allow you to eliminate operations that require setup. One company has started a vigorous program wherein their suppliers are asked for their ideas, to improve the products they supply, that would reduce the cost without reducing the quality. The initial results demonstrate that this will be a viable program with great success.

Simply displaying in your facility a part or product you either produce or use in your finished goods, and providing a form for people to submit ideas for improvement may provide great results. One company put a value analysis display in the lobby where supplier sales representatives wait when making calls on the purchasing department. It has received several cost saving ideas that have been implemented. Tap all the resources you possibly can to reduce cost while improving quality.

CONCURRENT ENGINEERING

Some companies have implemented concurrent engineering wherein they involve the assembly, manufacturing, design engineering, quality, manufacturing engineering, and the suppliers in the entire design process. This offers great potential to reduce set-up time while improving the quality of the product.

NEVER ASSUME

You have heard the saying "never assume" many times, and the principle applies to set-up time reduction. An aerospace company in Florida had a team working on set-up time reduction on a crush grinder. The operator/set-up person consistently said "that can't be done" to ideas presented at the team meetings. Almost as soon as the team began its work, however, their measurement graph showed improvement. They assumed that something was wrong with the measurement. They found out later, however, that while the operator was telling the team that he didn't think their ideas could be done, he would then discuss those ideas with the operators on other shifts and implement the changes without advising the team. One day when the team was making a visit to the machine in the shop, the operator was showing team members the changes he had implemented. Suddenly the team real-ized why the measurement graph was showing improve-ment. The operator said he assumed the team knew he was taking the ideas, discussing them with the other oper-ators and implementing the good ones. Technically he was doing what he had been trained to do—talking to the other employees—but he was not providing feedback to the team and getting the team's assistance in implementation. In addition, the improvements were not documented in the procedure. Once they were documented, the results were realized across all shifts.

COMMUNICATION AND COOPERATION

It is important that set-up time reduction teams communi-cate with the other employees in your company. Videotapes with documentation are very effective in demonstrating the new set-up procedure. Far too often, you see the lack of coop-eration or outright refusal to follow the new set-up proce-dure. Strong personalities who resist improvements from the team can impede those improvements. Getting input from as many set-up employees and operators as possible may overcome resistance. In some companies it simply comes down to the fact that everyone must follow the new proce-dure and supervision must deal with those who do not. If the new procedure is inadequate, the team must take respon-sibility for correction.

In addition to following the procedure, there may be a lack of confidence on the part of other employees who do setups. These employees may make changes on their own without working through the team. People have opinions that cause them to take actions and the set-up time reduction team must take the time to understand the opinions and either act on them, or eliminate the opinion with facts.

FLOOR LAYOUT

Establishing a staging area for raw material and delivering that material either while the previous job is running or early in the setup may greatly reduce the set-up time spent getting the material. Also establishing a staging area for first article parts of a new setup allows the quality department employees to know if there are set-up parts waiting approval.

One printing company found it beneficial to identify areas on the shop floor for staging of material. They started with a staging area for the sample load awaiting approval from quality inspection. Before, material was staged wherever there was room and the inspection department didn't always know what samples needed inspection. One food processing company established staging areas for raw materials that were located near the packaging line.

DATA ENTRY

In many companies, data entry during the setup is time consuming and a duplication of effort. For example, if your operator logs onto setup, the system should automatically determine when that run has completed. The system should not require additional data entry. In addition, you could implement bar codes to reduce the number of key strokes required to do data entry. Many companies today have a bar code on the employee badge that identifies the person and the job they do. Any efforts to reduce data entry will assist in set-up time reduction.

PAPERWORK

Much like data entry, the amount of paperwork required during or after a setup may be a waste of time. Have your set-up time reduction team identify the nonvalue-added paperwork and eliminate it as quickly as possible. Much of the paperwork done today is required because of a one-time problem or for "just in case." For example, duplicate entries into log books are made because the computer went down once and no one knew what jobs were running. Eliminate the cause of computer downtime and the paperwork becomes unnecessary.

Don't overlook approvals and signatures as a part of unnecessary paperwork. Anytime an approval or signature is required, it must add value, not double-check someone's work. If a signature provides communication, then it may be necessary depending on what the person does once he or she signs.

SCHEDULING

Most companies have found that communicating job schedules plays an important part in set-up time reduction. Use of scheduling boards ensures that everyone knows what job will be run next at each work center and allows the set-up employees to begin preparing, at least mentally, for their next job. Caution should be taken in determining how far out the schedule board goes. The next job is all an employee may need to know since schedules can change. If you go out too far, too much time may be spent changing the board and your ability to schedule properly may be criticized. If you do not allow for change, you will be disappointed since customer schedules do change—that is until you greatly reduce the set-up time. These scheduling boards can be placed in cells or departments as appropriate. Some companies have the set-up person identify completion of the setup by updating the board with a colored marker.

Another issue in scheduling is to determine which parts can run together with the minimum amount of setup. These are typically referred to as part families. The scheduler should then determine if any other parts in a "family" are needed when one of the family is scheduled. This is a good practice but when you get a great deal of set-up time eliminated, it may no longer be needed since you have quick setups.

TEAM STAFFING

If you have a steering committee, its job is to staff the teams, but it is important that the teams understand how the staffing was decided and why certain functional areas are represented on the team. If the team wants to discuss team staffing, members should be allowed to do so with the steering committee.

SUPERVISION

Many companies have recognized through set-up time reduction that the supervisors over the years have acquired responsibilities that simply aren't necessary. For example one company's supervisors were required to take the first article part to the inspection room. At some time in the past, an operator evidently had not delivered the parts to inspection and returned to his post quickly. Since then, the policy has been "only the supervisor can transport the first article parts to inspection." What a waste of time! This also contributed to the supervisor not being available in the work area. You should search out similar waste in your supervisor's time.

A common problem involves having the supervisor inspect the part after the set-up expert, prior to the inspection department performing first article inspection. Put a stop to this waste immediately. Once the set-up expert is satisfied, the part should be checked by inspection and there should be no need to involve the supervisor.

COMMITMENT

Measuring commitment is a difficult task. Employees are constantly questioning the commitment of management. Managers wonder if the employees really understand the importance of issues like set-up time reduction. If you conduct surveys of what employees and managers expect to see, in order to know there is commitment, here's what the typical responses will be.

We know there is commitment if employees and managers are:

- Breaking down barriers.
- Working together.
- Listening to suggestions.
- Nonthreatening.
- Providing positive reinforcement.
- Trusting.

This list comes from both employees and managers and it tells us that there is no substitute for time. Proving commitment takes time and effort; you must be in it for the long term and not give up. As soon as one side quits encouraging and participating, the initiative will die.

The measure of commitment usually involves money and time. If management is willing to invest their money into the effort, there is commitment. Likewise if employees are willing to invest their time in an effort, there is commitment. The president, vice presidents, and directors should all participate on set-up time reduction teams. Their participation demonstrates commitment. Likewise, as soon as you decide to start, money should be allocated to pay for the low-cost improvements that will be recommended. This makes it easy for the team to get the money and provides tangible proof of the commitment. I have yet to see a set-up time reduction team waste money. The team's frugality is not surprising since it consists of senior management, middle management as well as supervisors, operators, and set-up employees.

TRAINING

Many companies have recognized the importance of training all employees in set-up time reduction. This includes a short session for all employees, usually lasting two hours, whose purpose is for everyone in the company to understand the need for set-up time reduction. Other companies conduct a one-day training session for all employees to help them understand the subject well enough to be able to assist any set-up time reduction team.

Videotapes are excellent training tools for these sessions especially when the documentation is long or involves technical terminology. Dialogue that further explains the documentation helps. Some companies have used the video to show how to setup for a particularly difficult item, as well as for cleaning and maintenance.

CHECKLISTS

Team process checklists allow a team to evaluate their progress in development of team building. Each team member rates the team in each area and any ratings exceptionally high or low as compared to the rest are verbalized to create understanding.

PROCEDURES

Development of a makeready procedure at a printing company allowed better communication between shifts especially as improvements were implemented. Supervisors should ensure that the procedures are followed by all set-up employees and operators if appropriate.

NEWSLETTER

One company produces a monthly newsletter which consists of the accomplishments of each set-up time reduction team (22 at last count) so other teams can copy or use any ideas for improvement in their area. This newsletter can also be sent to other plants or divisions so they too can learn from other teams' accomplishments.

DECISIONS THAT AFFECT SET-UP TIME

Many management and supervisory employees make decisions that affect set-up time. Storing tooling in a central crib, kitting by part, lack of staffing in the tool crib during second and third shifts, and so on affect set-up time reduction. The teams must be allowed to investigate the decisions and work with management and supervision to resolve the effects of those decisions. Likewise, management and supervisors should be encouraged to work with the teams to reduce set-up time. Management should be aware of decisions that might affect set-up time reduction and seek the input of set-up time reduction team(s) that may be affected by a decision before it is made.

SUPPORT AREA RESPONSE

Much like the above recommendation, all support area employees need to be aware that their decisions may affect set-up time and that they should seek input from the team(s) if they determine that the change could affect setup. If a change is implemented that adversely affects set-up time, the team should be willing to reverse or alter the decision as quickly as possible. In addition, support areas should be willing to respond when asked for assistance by set-up time reduction teams. As long as the needs are reasonable, and are understood, the effort should be supported enthusiastically.

MATERIAL HANDLING

The installation of a hoist may not sound like much unless you are observing a setup. It is amazing how much lifting companies expect the set-up expert to do without assistance. Installing a lifting device may not greatly improve the set-up time, but it may reduce injuries and fatigue. Again, safety is of utmost importance when considering improvements.

At a glass manufacturing plant the set-up employees were having to move molds that weighed in excess of 100 pounds. Because of the limited area in which they had to work, they moved these molds while squatting and bending over. It isn't surprising that their backs were usually sore. As soon as savings are generated, you should consider investing in some material handling equipment that would reduce set-up time and make their jobs easier.

The set-up time reduction team should consider any material handling equipment and determine if it fits the application. Hoists may be the incorrect solution if air transfer is a better fix. Material can be moved on rollers, hydraulic lift tables, ball transfers, lift trucks, hoists, carts, or cranes. The team needs to evaluate all possibilities and determine the best method to implement.

CRITICISM

Although difficult for many to deal with, you should expect criticism of set-up time reduction initially. Some people generally find things to complain about and an effort like set-up time reduction may gain enough notoriety to become a target of criticism. Do not let that criticism stop the progress. Simply listen to it and learn from it. If there is no value to be gained, help the complainer to understand the fallacy of the criticism. Never take criticism personally and never attack the critical person. If you win over a critical person, they will be just as vocal in favor as they were opposed. All of a sudden, this negative is turned into a positive. Getting negative people on the first teams is another way to reduce criticism.

WHEN YOU CANNOT VIDEOTAPE

Initially in some companies it is not possible to videotape a setup. Companies with unions encounter this problem at first due to the lack of understanding how the video will be used. Time may overcome the issue like it did in one midwestern company. The union leadership assured me that there was no way a video would ever be taken of a setup. Two years later they agreed to allow the first video.

In most companies all you have to do is assure the employees that the video will not be used against anyone and then make sure that it isn't. If you absolutely cannot get a video, just have the set-up time reduction team observe the setup a number of times (five would be ideal) and take notes to document how to reduce the set-up time. This is how we had to do it before video equipment was available and it still works today. Use photographs to communicate and document in lieu of a video. Many times a series of photographs of a machine setup is a great help to the set-up employees and operators.

KANBAN INVENTORY METHOD

It is important for everything needed during the setup to be available and ready to use. Conventional methods of inventory management are not acceptable in some cases. Max-Min does not eliminate stock outs, due to the fact that lead times are not fixed as most systems require. Kanban is a good method of managing perishable tooling as well as production materials. Here are the steps of implementing Kanban:

1. Identify parts to be included (reoccurring usage items).

2. Identify maximum quantity to stock.
3. Identify *Kanban* quantity (standard container quantity).
4. Identify storage locations.
5. Allocate only enough area for maximum number of *Kanban* containers.
6. Design a frequent replenishment system.
7. Ensure system functions properly.

Once these steps are done, the material should be delivered in standard containers, frequently, and based only on usage. This method is applicable to all manufacturing companies, and if done properly, will eliminate stock outs. The key is the replenishment system that can deliver frequently (daily) in standard containers based on what was used. Kanban allows you to reduce inventory and eliminate stock outs.

MEASUREMENTS

Having measurements is vital to continuous improvement. You should never allow a team to improve anything until they have established how they will measure their progress. The team measurements must parallel the strategic goals and metrics of the company. Virtually every set-up time reduction team will measure time, based on this definition:

The elapsed time from the last part of the previous job until the first part of the next job is made at normal efficiency rate.

Beyond that, there are many possible measurements that could be used by your teams:

- Cost versus savings.
- Set-up time to lot size ratio.
- Set-up time to run time ratio.
- Work-in-progress reduction.
- Lead-time reduction.

- On-time delivery (to next operation).
- Average set-up time.
- Number of setups performed weekly.
- Cost of scrap weekly.
- Cost of rework weekly.
- Internal versus external set-up time.
- Set-up time per machine.

If your company has implemented cells, the following may be good measures of performance for each cell:

- Average part family cycle time.
- Average set-up time per machine.
- Throughput ratio.
- Product reject ratio.
- Inventory to output ratio.
- Average shop hours per part family.
- Average overload hours per month.
- Average number of parts to stores per cell employee.

Set-up time reduction efforts will contribute to improvement in all of the above areas. Never allow a team to have more than three measurements. One is minimum, but more than three and the team will spend too much time measuring and not enough time implementing change.

WORKFORCE REDUCTION

Reducing the workforce is very popular for most companies today. The shareholders see it as a positive move to eliminate costs. This is only because there is no line item in the financial statement that shows the cost-to-carry inventory and the cost of quality. If there were, the shareholders would see that there are greater opportunities in these areas. Avoid reduction in labor as a result of set-up

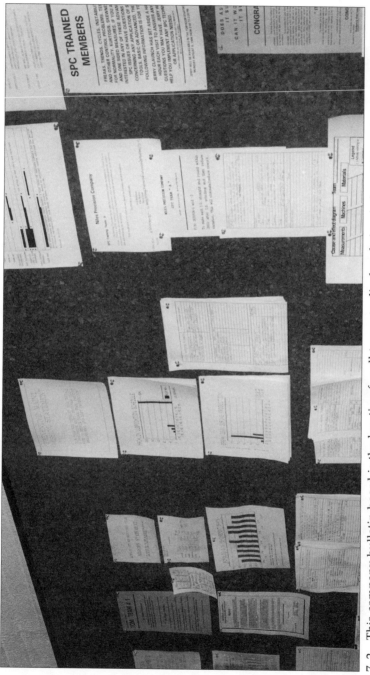

7-2 This company bulletin board is the location for all teams to display their mission statements, cause and effect diagrams, and current progress measures.

7–3　This shaft manufacturing cell bulletin board provides a place for the team in this cell to display their mission statements, cause and effect diagrams, and current progress measures.

time reduction. Let's face it, if you ask employees to be on an improvement team and layoffs occur due to the improvement, the remaining employees will be reluctant at best to continue the improvement effort. If any employees wind up with no work to do, put them full time on set-up time reduction.

Most companies have the financial department identify the value of one hour of additional production time. This then allows the company to focus on set-up time reduction, not labor reduction. In addition, have the finance department identify the formal waste, scrap, and rework costs. The true cost of quality will be from 800 to 1,000 percent greater, but it is very difficult to track and measure.

One printing company identified their waste costs to be $4 million annually. The set-up time reduction teams were trained to use the problem-solving method and to identify quality problems. At the end of the second year, waste was reduced to $3.25 million while sales had increased 66 percent. Their make ready reduction teams had reduced cycle time thereby allowing for more sales while the amount of waste was reduced. As you can see, the true benefit was much greater than $750 thousand of waste, especially considering that their sales had increased as well.

SPECIAL TEAMS: SWAT AND RAPID RESPONSE

Special Work Assignment Teams (SWAT) have been used to attack set-up time by some companies. These teams become very proficient at achieving set-up time reductions and are usually joined by the set-up employees and operators from the work area.

Other companies have organized a set-up reduction task force to quickly address and act on ideas to reduce set-up time. The team members need to have excellent people skills and should be kept small. The core team usually consists of the manufacturing engineer, the supervisor, and the tool designer.

When joined by two set-up employees and an operator from the work area, the team becomes a normal size, but the results can be accomplished quickly since they become expert at the task. The SWAT team takes leadership and develops an understanding of the current setups and changes that can be implemented with the new members quickly.

When the SWAT team is finished it is important that a set-up procedure be provided to the work area in order to ensure that the improvements are followed in the future. You will also find that a videotape after the improvements are made is helpful in training of the new procedure. The important thing is to provide measurements that ensure the changes result in reduced set-up time.

Rapid response teams are also employed by some companies to respond to quality problems that are encountered during setup. The operator or set-up reduction team can activate the rapid response team when needed. Normally the supervisor and quality assurance specialist are included in rapid response teams, along with manufacturing engineering, the operator, and set-up expert.

QUESTIONNAIRES

Many companies have found questionnaires important to determine the status of set-up time reduction. Included in this book is a questionnaire that may assist you in the development of your set-up time reduction program (see Chapter 10, page 162).

If set-up time reduction is to become a reality in your company, it must be properly supported and encouraged. Although you expect results from the team, improvements may stop and the initiative fail completely due to lack of support. Don't let that happen in your organization.

CHAPTER 8

Holding the Gains

One key to an effective set-up time reduction effort is to hold the gains made. It is important that the improvements made continue to be used, especially as the team moves to another machine or area of the plant. This will require a change in methods utilized by many companies due to the fact that setup is not documented nor is a procedure developed. Most set-up experts have developed methods and techniques, yet there is little communication or sharing of that knowledge among other set-up employees.

DOCUMENTATION

The first step is to provide a complete documentation of the setup, but how that documentation is presented is also important. This documentation is significant because it will provide an accurate documentation of the steps as well as important considerations during the setup. The easier it is for the set-up expert to visualize the steps, the easier the steps will be to implement. One of the keys to achieving set-up time reduction is fully implementing the solutions. Set-up employees on all shifts must follow the same set-up method. This is difficult or maybe even impossible without documented procedures which could be a checklist or a routing with detailed instructions. Whether simple or complex, in all cases instructions are necessary. A videotape of a solution is an excellent training tool when coupled with a checklist to be used by the set-up expert.

These procedures need to be as simple as possible and the easiest solution may be to add pictures to the procedures. In most companies, procedures contain no pictures. Words are boring and difficult to understand. Providing pictures will enhance understanding, eliminate thousands of unnecessary words, and get people using the new procedure quickly.

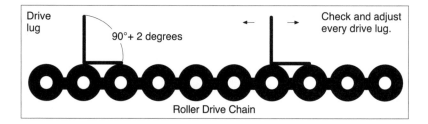

This picture was developed by a set-up time reduction team at a company in the northeast part of the United States. The team found that a great deal of set-up time was spent getting the boxes square after the product had been inserted, and just prior to the application of the heat seal. After they concluded their root cause analysis, they determined that the angle of the drive lugs was critical. They provided a snapshot of the situation to the set-up expert, and he used a red marker to identify the 90 degree critical dimension to be verified.

The team then met with the supplier to develop a stronger drive lug so this problem would not happen in the future. This picture was all the set-up employees needed to communicate the change and was put in the procedure manual. Without the picture, this set-up procedure would have required a lot of writing and reading to make sure everyone knew what to check and what to fix. Encourage your teams to develop methods of communicating procedure changes effectively and quickly. Pictures are well worth consideration, especially with the advent of digital cameras that are compatible with computer-generated documents and scanners.

The people receiving the documentation must follow it and provide feedback to the team as problems and solutions arise. There may be a competitive culture in your facility in which feedback and cooperation are the norm. If this is so, you will need to think about how to provide the documentation, who will receive it, and how to ensure that it is followed. Your steering committee should work with supervision and the set-up experts to determine how this will be done. This process needs to be in place when the first set-up time reduction team is prepared to implement changes on their first machine.

The documentation of the set-up is an important task that the team should not take lightly. Once a video is taken, the team documents the current set-up practice, which the team intends to improve. As the team decides on changes, the documentation must change. From here on out, this document will instruct the set-up employees in the new method and sequence they are to follow in order to accomplish a correct setup. It is important then that the team provide an accurate document that is usable and understood by those employees.

Experience has demonstrated that the team members understand the changes they have made, but they do not always communicate them to supervision or other set-up employees properly. This communication is the responsibility of the team and is easiest to accomplish if there is a standardized format for communicating the improvements. The purpose is to reduce the set-up time, and without standardization the previous waste will continue.

CELEBRATE IMPROVEMENT BY MAKING IT A BIG DEAL

By making a lot of commotion about an improvement, you heighten the awareness of change. When a team makes significant improvements to a setup, the entire organization should be made aware of what has happened. Set-up time reduction is no small accomplishment—you want everyone to recognize the improvement. The other set-up employees will need to change how they do that setup. Simply putting a notice on the bulletin board may not be enough; employee briefings may be in order. You have got to make sure that everyone understands the importance of making improvement and, once made, holding the gains. Normally a company creates excitement at the beginning of an initiative like this, but the time to make that excitement a reality is when improvements are implemented.

STEERING COMMITTEE REVIEWS

The steering committee should conduct reviews of the team progress. These should be at regular intervals and should not take a great deal of time. Five to 10 minute presentations to the steering committee on a monthly basis should be sufficient. The teams should present the following:

1. Current measurement graph(s).
2. Decisions made since the last presentation.
3. Changes implemented since last presentation.

The steering committee should not second-guess the decisions of the team, they should only be concerned as to whether the team is achieving their goal and using the proper method to reduce set-up time. If the team is doing well, the steering committee should tell them so and encourage them to keep it up. If they are not doing well, if they have gotten off track or otherwise are in need of help, the steering committee should provide that assistance and get the team on track. The steering committee should ensure that set-up time reduction teams are properly trained in the tools to be used for set-up time reduction, and that team leaders are trained in their role as well. Team recorders should have a set format for taking meeting minutes that provides an easy method to determine if the team is on track and progressing.

FIELD TRIPS TO THE FACTORY

Encourage the steering committee, other teams, and all employees to make scheduled trips to the area where the improvement is happening. You may listen to a presentation, but nothing replaces seeing the change and discussing it first-hand with the people involved

Without fail, these trips into the shop or office area generate feedback that is positive, exciting, and community building—everyone becomes better informed. You can listen to a presentation, but true excitement happens when you experience it first hand. On one occasion, a computer numerically controlled (CNC) lathe set-up time reduction team made a presentation to the steering committee and discussed how they had organized all the tooling, shim stock, face plates, and fixtures. The team leader invited the steering committee to come out to the area for 10 minutes to see what had been accomplished. During the presentation, the steering committee was complimentary on the accomplishments. After the field trip, the senior managers wouldn't stop talking about the improvement. We've been trying to do that for years was a common phrase. This team had organized a department meeting to get input on how the items should be organized and then did it. What had been needed for years is now a "done deal." These field trips will stimulate excitement and thereby possibly prevent the initiative from becoming a passing fad. Excitement may be the ingredient that keeps set-up time reduction from failing.

TRANSFERABILITY (CLONING)

Every implementation made by a team has the possibility of application in other areas, departments, plants, or divisions. The steering committee should develop and implement a method of communication between these business units. The steering committee should take the responsibility to have the team communicate in such a manner that transferability is made possible. Joint presentations, videotapes, and field trips are all possibilities. If your plants compete with each other for the product to produce, a method to overcome that concern will need to be developed. We must learn from other achievements within the company.

REWARDS, AWARDS, AND OTHER MOTIVATIONS

Rewards are very important in any improvement effort, and set-up time reduction is no exception. There is a great deal of effort that will be expended and rewards encourage a continued effort. Most companies simply ask the team what they would like to receive in the form of a reward and normally it is something that will help them on their job, not some personal gain.

Any time the issue of rewards is discussed, financial awards always comes up. You should consider not offering financial awards, since they seem to cause problems. Let me explain.

I know of four large Fortune 500 companies that offer financial awards to teams based on the savings. In all four the problems are the same and no one seems happy with the system because:

- Many people not on the team were instrumental in the implementation and they don't receive any money.
- People that drop off the team don't receive money.

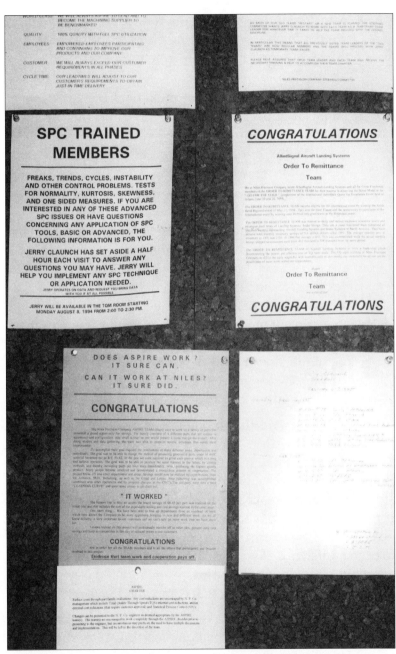

8–1   This manufacturing company has a bulletin board section devoted to recognizing the good work of its teams.

- After the initial set-up time reduction, the financial impact may be less and the effort stops because the team won't receive as much money in the future.
- There is a committee that must evaluate the financial impact of the team effort and they are inundated with presentations that show savings. A great deal of effort is required by many employees to sort out the savings.
- The facilitators are not team members and typically don't receive any financial awards. At one company, a set-up time reduction facilitator had 14 teams, 7 of which received the highest financial award of $1,000 per team member, 3 received $750 per team member, and the remaining 4 teams received $500 per team member. The facilitator's comment to me was, "I didn't even get lunch when the awards were passed out." Fortunately this facilitator doesn't get distracted, and is still leading the effort with great success, but it is not fair that he is left out of the award system.

It has been said that money does not motivate, and is in fact a demotivator. No one is going to turn down any money that is offered, but there are real ways to reward effort that don't cost much and are motivational. On the positive side of money, many companies have had great success with companywide profit sharing or gain sharing programs with no demotivating effects. The companywide application seems to be the difference. If you want to reward with money, include everyone in the reward system.

Here are some rewards that companies have offered their team members:

- A "wall of fame" whereon a certificate is placed for every team that achieves a 50 percent reduction in set-up time with all members' names on the certificate. One company includes pictures of the team as well.

- A brass plaque at the machine with team members' names on it when the set-up time is reduced by 50 percent.

- Recognition of teams in the company newsletter with pictures.

- Recognition of teams in a local newspaper. (Imagine the sense of pride when a neighbor or friend calls and says I saw your picture in the newspaper yesterday.)

- Certificates presented by the company president or chairman of the board.

- Dinner for two for everyone involved in the reduction of set-up time.

- Companywide luncheon served by senior management in appreciation of set-up time reduction efforts.

If teamwork and continuous improvement are to progress, the rewards must not segregate the performance levels. If there truly is a team of the month, then that means that there were failures; I don't share that opinion. It seems to me that a reward system that recognizes the improvement and creates the desire on the part of the team to accomplish more in the future should be the goal.

One company offered 10 percent of the savings to the teams as a reward for their efforts. About the third year, the amount of money given as rewards was becoming quite large and the company was reluctant to continue the system. The problem was that the benefits were not showing as an improvement in profits. The owners decided to discontinue the profit sharing and when the financial award system was taken away, the team process failed as well.

One company began its set-up time reduction improvement when it was in the throws of chapter 11 bankruptcy reorganization. After two years the company had improved to the point where it was able to put the reorganization in place and decided to reward all the employees with a monetary bonus based on one-third of the improvement profits. Many of the employees asked if they could return the money because having a job with security was more important than this one monetary bonus. They thought that this bonus might put the company back into chapter 11. The company management would have never offered the bonus if it would put the company back into financial difficulty. In fact that company is doing quite well today and is in great financial position.

There are some financial award systems that work, because the award is linked directly to actual performance. Spend time listening to the people who manage the reward system; then spend time finding out firsthand from current and former team members how well the reward system works. Many people are simply not motivated by money. They will take all the money you give them, but it doesn't motivate them to keep doing set-up time reduction or achieve the levels of performance you intend. Be careful if you intend to remove existing rewards, but becoming creative in the future reward systems may offer greater returns.

Ask the Team Members

You could ask the team members what they would like as a reward for improvement. Probably without fail, the teams you ask will respond with something that helps them do better in the future. A special tool, a storage rack, music in the work area (be careful here, it's hard to meet everyone's taste), an upgrade in the computer system, digital readout, and so on. The real key is to determine what motivates people in the first place. If they want money as an award, then base that on performance and be consistent in applying it to all improvement teams.

Recognition for a Job Well Done

Recognition for a job well done motivates people, and the recognition may be very simple to give. Allowing the entire team to participate in a presentation to top management can be a real boost to morale. They won't stop improving; they will want to do even better next time. By top management, we're talking about the chairman of the board, chief financial officer, chief operations officer and president. The teams may need help in preparing the presentation and the top management should show genuine interest—ask questions, and verbally recognize the good job. Recognition should also be accurate. Top management should not tell any team they are the best. They should, though, tell them what kind of job they have done. In my opinion, any improvement that helps your company is excellent. All these achievements will add up to becoming world class.

Plaques and Certificates

Plaques and certificates are an excellent way to recognize effort. They don't have to be expensive; they just need to be sincere. One company in the southeast puts a plaque at the machine every time a team reduces the set-up time by 50 percent. On that plaque are the names of all the team members. Another company gives a certificate to every team member when he or she reaches certain milestones. One company has a "Wall of Fame" where a certificate is posted when a 50 percent reduction in set-up time is achieved. The certificate contains the names of all team members with a team picture. This is an inexpensive way to make people feel good about their effort.

Newspapers

Consider advertising in your local newspaper the achievements of your teams, especially if you live in a community of 250,000 people or less. When your teams achieve a certain level, recognize that effort in the newspaper with names and pictures. Don't do it once a year, do it as the team(s) achieve their goals. Likewise, don't include a lot of teams and names in one article. A quick, easy-to-read piece with the names predominant will suffice and provide motivation.

Management Presentations to the Team

Another way to recognize effort and motivate for the future is for management to make presentations to the team. Present exactly what the result of the team's work has been and how it helped the company in sales, efficiency, or customer satisfaction. This will encourage the team to do even more in the future. These presentations should be specific and directed toward achieved improvement. Generic presentations to many teams at once may not motivate them at all.

Letters to Personnel Files

Some companies have found that letters that are copied into an employee's personnel file are a good way to motivate as well. I still remember the letter from a company vice president recognizing extra effort on my part that was copied to my personnel file. There may be nothing better than knowing that the recognition will not be forgotten. When it is time for promotion, or evaluation, that letter will be there reminding my boss of what the company recognized as a benefit.

Sincere and Specific Recognition

It is important that recognition be sincere and specific. Any recognition given halfheartedly, to a large group, without specific accomplishments cited, does little if anything to motivate. If it was great, say so. Set-up time reduction teams should hear:

- "That move to external was sensational."
- "That improvement in standardized tooling was great."
- "This team implemented something we haven't been able to do in the past."
- "This customer was kept because of the job this team did."

Look them in the eye, tell them they did a good job, and tell them you appreciate it. They'll do even better in the future and you will feel a part of the process as well. When it comes to recognition, two things are very important. First, you must not let the benefits gained be lost in the future, and second, you need to find ways to recognize the efforts. Properly done, this will serve to motivate employees to future gains and achievements that exceed your expectations.

CONTINUOUS IMPROVEMENT

As the name implies, continuous improvement never stops; it is continuous in that ideas build and better ideas may come along. You will do well if your teams never feel they have come to the end. Some ideas implemented today will need to be replaced with better ones in the future. Your teams should not strive, however, to find the absolute perfect solution. If they search for perfection, they will delay the process of making continuous improvements. Spending a reasonable amount of time making steady improvement while working for new and better ideas regularly is the way to go. If you have improvement happening often and that improvement never stops, you will have great success.

CHAPTER 9

What Should You Expect from Set-Up Time Reduction?

RESISTANCE AT FIRST

It is normal for employees, especially the set-up employees, to resist initially the idea of set-up time reduction. They have accepted set-up time as normal for many years and now you are asking them to help reduce that time. Its importance, and how to accomplish it, may not be understood in the beginning. Don't be overly concerned. Those who resist will become the best supporters once they see that set-up time reduction works and that there is commitment from the company. The training you give should provide understanding of the need for set-up time reduction, enthusiasm for the subject, and the know-how to accomplish the results. Providing examples that the employees can relate to are necessary as well.

FREQUENT AND CONTINUED IMPROVEMENTS

You should not be satisfied with the initial improvements; you should expect continued improvements frequently and over a long period of time. Frequent improvements on a continuing basis are what will keep your organization world class.

A television commercial says it all, "you can't become the best by comparing yourself to the best." You need to identify the improvement areas; then make sure that the improvement never stops. If you demand that improvements be implemented while respecting the employees, you will create the right environment. Pressing continually for results is the environment that is needed.

BREAKING DOWN BARRIERS (TEAM BUILDING)

It is important that you give set-up time reduction the opportunity to develop and become a way of life. At the start, your efforts will be geared to breaking down the barriers that have developed over time. The amount of initial improvement in set-up time reduction may be small compared to what you had planned. As the team develops and gets through its initial stages of growth, it is developing skills that will yield a great deal of change later on. The first stage may last a few months and you may only realize a breakeven in the cost versus benefit analysis. Don't give up or expect too much, but support and press for results and the improvements will come.

COMMITMENT

The amount of employee commitment initially may be barely enough to get started, but as set-up time reduction becomes a way of life, the demonstrated commitment should grow. You must prove to some employees that the organization is truly committed and realize that, until time passes, they may doubt that commitment.

"Flavor of the month" is a term we often hear in respect to past programs organizations have tried. You will do your organization a favor to fully analyze any new initiative and determine if it will take the organization in the needed direction. Don't start set-up time reduction unless you are personally going to demonstrate continued commitment over a long period of time. Otherwise set-up time reduction may become one of those programs that just came and went.

STAGES OF EMPLOYEE ATTITUDE

Just as teams have stages, so also do employees with respect to their attitudes. Keep in mind, those that jump on a program early must be supported and encouraged. Remember as well, those that are initially against an initiative may become your best supporters later on.

It is not unusual for employees to be very vocal about the fact that they are dead set against set-up time reduction. Some may be so adamant that they become offensive to your trainers. These are not bad employees. All you need to do is get them to understand the need for set-up time reduction and present a proven method for achieving it during the training. At the end of the training, they should be willing to give set-up time reduction a chance. Many times employees will do a complete turnaround and become your best supporters.

Examples are plentiful of employees who were against the effort at first, but once they saw the improvement possibilities and were certain that the support was in place, they became the best supporters. Employee attitudes will change and go through stages as they see the benefits, implement the changes, and become certain of the commitment of management.

MANAGEMENT STAGES AND ATTITUDES

Senior management is normally the driving force that starts out totally in favor of set-up time reduction. They are vocal about the initiative and encourage everyone to get involved. However, if they stop talking about set-up time reduction, the initiative may die out. It is easy to start; the difficult part is finishing. A resolve to get the job done is the only way to begin an effort such as set-up time reduction. Senior management must recognize that they are making a commitment for many years, not just a few months. Set-up time reduction takes time, and it is not something that can be completed quickly. If you are looking for a quick solution, look elsewhere. If you are looking for one of the key initiatives that will get your company to world class and keep you there, you have found it. Senior management should decide if they are going to remain committed, and participate over the long haul. Once they answer yes, it is time then to get the rest of the organization involved.

Middle managers and supervisors may be slow to accept the initiative because, at times, it will interfere with daily pressures such as getting shipments out. Set-up time reduction will make these daily pressures go away, but it takes time for these people to grasp the subject and see it to completion. It takes time for results. Once they are committed, the initiative will have strong roots and should achieve noticeable change. In your new environment, your managers continue to have a great deal of pressure to get daily tasks done, while at the same time reducing set-up time. Once they begin to see the improvements made possible by set-up time reduction, they will press for results and the commitment will be demonstrated.

Future expectation of greater change will be a way of life once each level of management accepts its responsibility and commitment to the improvement. There is a big difference between allowing change to happen and expecting great change. When you expect great change you push daily for improvement and once you arrive there, the improvement will not stop. Prior to that, the improvement is susceptible to discontinuance at any time and therefore is fragile.

PRODUCE AS NEEDED

Two benefits of set-up time reduction are increased capacity and inventory reduction. Do not look for labor reduction as a result of set-up time reduction. Labor reduction today is heralded by Wall Street as companies slim down. Soon, the experienced labor that has made our companies great will not be available. Companies need employees, as well as cost reductions. The increased capacity that allows you to produce as needed holds far greater benefit in the long run. As set-up time reduction reduces the set-up time, you must push for increased capacity. This increased capacity should only be used, however, if you have demand for product. As this happens, you will find less need to hold finished goods and work-in-process inventory. The cost reduction benefits due to inventory reduction, will allow you to satisfy the shareholders with the increased profit they deserve while maintaining the employment level.

INCREASED RUN TIME (BUT BE CAREFUL)

If the desired result of set-up time reduction is increased run time because you find that you are constantly out of capacity, or late in delivering to your customers, increased run time will be a benefit you realize. Be careful not to just increase run time when there is no need. If you are not out of capacity or don't need additional production to meet delivery, let the machine sit available rather than increase inventory. Too many companies miss the importance of inventory management in cost reduction. If increasing production (run time) increases inventory, then it was unnecessary and of no benefit.

Your finance department may present figures that convince you that a machine that produces more, costs less. This is only true if you sell more product. Take another look at the formula for productivity:

Productivity = $ of Output (divided by) $ of Input

Output only occurs when a product is invoiced or, more technically, when the invoice is paid. If you produce more but it goes into inventory, you have increased the input costs. In this case, your set-up time reduction doesn't provide a benefit. Soon everyone will be ready to stop this initiative.

Producing the same amount of product in a shorter period of time will allow you to reduce inventory. This provides less input costs with the same output and thus higher profits. Examine your demand, examine your costs, and determine how to increase profits; then you will be on the right track. In most companies an idle machine provides a number of benefits such as:

Preventive maintenance time.

Time to work on scrap reduction.

Time to work on rework reduction.

Cost avoidance.

Training time.

Team meeting time.

PREVENTIVE MAINTENANCE TIME

Far too often companies do not allow time for equipment breakdowns to be prevented. Maintenance is competing with production for the time to perform the proper repairs to the equipment. Preventive maintenance is essential to a world-class company to keeps costs low while providing on-time deliveries to its customers.

Set-up time reduction may provide just the needed time to perform proper machine maintenance in a preventive mode. Everyone gets what they need, including the maintenance department which gets the time it needs to perform proper maintenance. Freeing up the hidden capacity allows you to make the right decisions about how to use the additional capacity, but don't just jump to the conclusion that you need to increase production. A well-maintained machine is worth its weight in gold to a world-class company.

SCRAP REDUCTION

As you charter the set-up time reduction teams, you must stress that scrap during setup is not tolerable. If scrap presently occurs, the team must find the cause of the scrap and eliminate it at the lowest cost possible. Often, scrap is simply accepted. It has always been there; therefore it will continue. If that is the case, set-up time reduction teams must recognize that scrap is the first improvement that must be made. The steps are:

1. Eliminate the cause of scrap.
2. Reduce the set-up time.
3. Set up frequently.

Scrap is the enemy and its causes must be attacked and eliminated by the team. Any scrap is a waste of a company's resources, its labor, material, and time. Thus, scrap reduces the profit potential of the organization!

REWORK REDUCTION

Much like scrap, rework must also be identified and eliminated by the set-up time reduction team. When you see rework charged to set-up pieces, the team must address that issue before it increases the number of setups. Rework, like scrap, is a quality issue that must not be tolerated. The team must find the cause of the rework and eliminate it using the lowest cost possible.

Be aware that the rework done may be informal. It is not reported; it is just accepted as a normal step in the process. Don't accept that. If rework is currently occurring during setup, eliminate it.

Cost Avoidance

Some companies don't recognize cost avoidance as a benefit because it would be easy to claim cost avoidance when someone was simply doing their job. Don't criticize this practice of recognizing cost avoidance. With set-up time reduction, you may in fact realize cost avoidance in four possible areas:

1. Equipment purchases.
2. Additional production shifts.
3. Additional inventory for increased demand.
4. Adding inventory for production equipment downtime.

EQUIPMENT PURCHASES

If your organization has determined that additional equipment must be purchased in order to keep up with demand, an examination of the benefits of set-up time reduction is in order. If a reduction in the set-up time will allow you to eliminate the need to purchase the additional equipment, then that cost avoidance can be credited to set-up time reduction once achieved. The same may be true if you plan to add additional production shifts to keep up with demand. The examination of set-up time may indicate that this cost can be avoided if a reasonable amount of set-up time is reduced.

REDUCED INVENTORY

One company found that reducing the set-up time completely eliminated the need for the additional warehouse space. They were in the process of building an additional storage facility. Once they proved that set-up time reduction would work for them, they stopped the building immediately even though the footings were in place. All other costs were avoided. You should also examine if set-up time reduction and the subsequent smaller lot size production (make to order) would allow better utilization of the current facilities. This may allow for discontinuation of present or future building plans either for production or warehousing of inventory. There may be a lot more costs that can be avoided if set-up time reduction were achieved, such as:

Tooling expenditures.

Emergency maintenance.

Wasted time looking for change parts.

PERFORMANCE CHARTING

Every company that decides to implement set-up time reduction should insist on performance charting for a number of reasons. First, performance charting allows the team to monitor its own performance. This allows team members to determine if their performance is adequate or needs improvement. Second, it allows the rest of the organization to see the improvement and monitor the progress of all the teams. If one team is having difficulty, the steering committee or facilitators can provide encouragement or assistance as necessary. Third, performance monitoring provides an overall view of the progress made and the improvement the initiative provides to the company as a whole.

Any team chartered to make improvements must have performance measurements. Without measurements, teams may become discouraged and stop. With measurements, the team members can determine if they are satisfied with their progress or if they need to move to higher levels of accomplishments. It is best if the performance of a team can be measured in productivity improvement. Everyone should be encouraged to monitor the team activities. Don't pass up the opportunity to encourage the team effort.

PRODUCTIVITY IMPROVEMENTS

Earlier in this chapter, we examined the formula for productivity. If your teams are not providing productivity improvement, something is wrong. Productivity must increase in order to avoid having to reduce employment as a method of increasing profits. These productivity improvements must contribute to profits to be viable. In most cases it takes the efforts of more than one team to provide the overall productivity improvements needed. Therefore it becomes important that the organization track and realize productivity improvement as a result of the teams' efforts.

How you measure the productivity is as important as the measurement itself. Involve the finance department to assist in the calculation of the measurement. To capture the cost of goods sold, in its simplest form, you subtract profit from the annual sales. This should suffice unless there are substantial reserves being used or accrued, or other sources of income to the organization such as lease or by-product sales.

The measurement should be kept simple so that reporting is timely and accurate. This productivity measurement should be posted for everyone in the organization to watch, since providing these reports for everyone in the organization to see encourages continuous improvement. You should question any information that is withheld and eliminate the withholding as much as possible. The more employees know about the performance of the company, the more they will be willing to contribute to its success.

TEAM BUILDING

The building of teams and the team networks is certainly worth the company's effort and time. Teams work together and provide greater results than individuals, but teams take time to develop. Be patient and allow the normal stages of team development to occur. This development allows the teams to perform effectively at a high level of achievement.

Team building will generate an environment where decisions are not made by individuals. This is important because most individuals can only see a problem from his or her point of view and a solution from his or her background. Teams surround the problem with understanding, and when problem solving is applied, team members implement solutions that eliminate the problem. Individual problem solving because it is examined by one person's point of view usually masks the problem rather than eliminates it.

I'm sure you have been in environments where fire fighting is the way problems are addressed. The result is many partially addressed problems that do not go away. The employees are busily providing information, reports, and frantically trying to get shipments and quotations completed for customers; the place is chaotic. The solution may be to computerize, but then you still have chaos, only faster. The problem is not the people or the system, it is that you are not allowing for true problem solving. The symptom not the cause is being addressed.

In a fire-fighting environment, supervisors and managers are very aggressive and the rest of the organization relies on them (or more accurately, stays out of the path of destruction) to get the job done. Frustration is great, and so is conflict, but at the end of the day those supervisors and managers have a great sense of pride in getting the job done. The rest of the employees simply go home and dread tomorrow. Probably the best analogy is that these supervisors and managers are outstanding individual players in what should be a team sport.

For years, the Negro College Fund has aired a commercial that says it all: *A mind is a terrible thing to waste.* Every employee has a mind, each has experiences to build on, and each can help to solve problems. All you have to do is let them. Teach them problem solving, train them in set-up time reduction, let them work every week as a team and be prepared for tremendous results. Don't be concerned if they resist at first. This is new territory and they may be uncomfortable with this team environment, knowing how the organization has run in the past.

BREAKING DOWN OF BARRIERS

Breaking down barriers is what teams do for an organization. Again, set-up time reduction will take time, but soon people will realize that they can contribute and there are no barriers if they respect the other people and communicate effectively. Remember: the key to effective communication is listening.

In the team meetings, you should stress that one person speaks at a time and no one overspeaks. Here is a good rule-of-thumb formula that ensures equal participation at team meetings:

1 (person) divided by (total number of team members) times 60 (minutes of meeting time) = how long each member should speak at each team meeting.

If you had 7 people on the team, the formula would look like this:

1 divided by 7 = .14 times 60 = 8.4 minutes.

Each team member should be encouraged by all the other team members to give their input. When one speaks, the others listen. Everyone participates, but no one monopolizes the conversation.

Sometimes one member will talk too much at first. Usually as time goes on, the participation by the other team members will begin to equalize; they all learn to give input some of the time and listen to other input most of the time. Soon you will find that the barriers to ideas and questions are replaced by receptivity.

The physical barriers will begin to disappear as well. Factory employees won't be stopped at the doors to the office area, because they have needed to go through those doors to get information or to solicit help from department employees for set-up time reduction. Office employees will be seen in the plant more than they used to be and they will begin to understand the production process as well as the problems encountered.

One set-up time reduction team at a company in the Midwest was frustrated at first because they had a team member from the finance department. They complained that they were moving slowly because they had to explain so much to this employee who had never worked in production. About three months later, they were elated because one of the solutions required a small amount of money and this finance employee got the approval with one phone call. Soon this group was functioning as a fully empowered team. The key was that they recognized the strengths of each team member and overcame the weaknesses. Barriers will go down in this type of environment. Finance employees, factory employees, managers, and supervisors are all very smart people, but they don't know everything. Teams can solve anything as long as their members don't stop at the barriers.

INVENTORY TURNS

Since reduction of inventory with no stock outs should be one of the primary benefits you receive from set-up time reduction, make to order should be your ultimate goal instead of making to stock. The measurement of inventory turns should be calculated for the entire organization since it frees up working capital as well.

The formula for inventory turns is:

Forecasted cost of goods sold (next 12 months)
divided by total inventory

If your forecasted cost of goods sold is $12 million and your total inventory is $1 million, your inventory turns are 12.

With 12 turns, you would be better than most manu-facturing companies who measure inventory turns, but you should never be satisfied with the current status. Likewise if your inventory turns are low (three to six) don't become discouraged, simply begin set-up time reduction and drive the inventory to a lower level. Whether sales are increasing or decreasing, inventory reduction is valuable to the profitability of the company. (You can control purchased material inventory to a certain point; then your suppliers must implement set-up time reduction for you to get greater results.) You do have control over two inventory classifications: work-in-process and finished goods.

Work-in-Process Inventory

Work-in-process (WIP) inventory normally accounts for the most significant inventory companies have on hand. It is costly and sits most of the time in queue waiting to be run. WIP may be the driving force in your ability to compete and certainly has inherent problems of its own. If you take the step of reducing the set-up time, and then reducing the lot size, you will find that WIP inventory reduces automat-ically. It is important to recognize that you will need to reduce the set-up time on many machines in order to reduce the lot size, but once you see the result, you will push for those results in the future. *This is where you may do some "what if" scenarios:*

What if your WIP inventory were one-half of what it is today?

What if your economic lot size were one-half of what it is today?

What if you could cycle through all the different parts, flavors, or items you produce in half the current time?

What if you could reduce your cycle time by one-half?

All the above can be accomplished by simply reducing the set-up time and then reducing the lot size.

I'm sure there are people reading this section saying they would never be satisfied with only one-half reduction in the above scenarios. Neither would I, but without some milestones, you may never realize the reduction. Suppose you tell a set-up time reduction team in a cell that you want startling results, 90 percent minimum. Your product goes across six machines before it is ready for shipment. The team works on one machine until the 90 percent minimum reduction is achieved. You may have missed the opportunity to reduce the lot size because now one of the other five machines is holding up the reduction.

Overall results will only occur if the entire process is reviewed and reduced, not if only one part of the process is reduced. If you push for a 50 percent reduction over and over again on all machines in a process, and never stop, you will realize greater results faster.

Finished Goods—Total Supply Chain

A number of company managers say they don't stock finished goods; they make to order, but they always seem to have finished goods. Responses like that's not ours, the customer has paid for it are incorrect.

These managers feel that it is acceptable to have inventory as long as the customer has paid for it. The customer paid for it, but it is still at your location. Does the customer really need the product? What if your competitor called the customer and said, "You no longer have to forecast out three months. We've reduced our lead time through set-up time reduction and accomplished lot size reduction, so you only need to forecast your needs for six weeks." You probably would lose a customer.

It doesn't matter how you account for the inventory financially. Inventory is an indicator of opportunity to reduce costs either to you or to your customer and both are important. If you have inventory, see it for what it is—an opportunity.

If you make to stock because the customer is always giving short lead time orders, then reduce your lot size by reducing set-up time, and your lead times will reduce as well. When your lead times match the customer's order cycle, you can eliminate finished goods completely and do the ultimate—make to order.

SPACE

Space will become available primarily due to the reduction of WIP inventory. In many companies, when managers start set-up time reduction, they don't realize how much space is tied up due to the material in queue, or scattered around the facility, or located in the storeroom, partially finished. With lot size reduction the result is less WIP inventory. Examine the inventory that is no longer required. Get the set-up time reduced; then run smaller lots. Push hard to get the lot sizes to the quantity your customers require and your lead times to match the customer buying habits.

FLEXIBILITY

Being flexible means being able to respond quickly. This is key in the future of world-class manufacturing. Quick setups coupled with small lot sizes will make your organization more nimble. Produce the product quickly, ship it, and invoice with no leftover inventory and the future belongs to you. Until product is shipped, all kinds of bad things can happen to it, including damage, loss, pilferage, and counting. In addition the organization incurs costs—storage costs, tracking system costs, cost of funds, equipment costs, and so forth. Now add engineering changes which create obsolete inventory into the equation and the results become overwhelming.

Set-up time reduction allows you to become flexible and realize the benefits of lean manufacturing. The competition will have to meet your standard of responsiveness while you continue to become even more responsive with the continued results from set-up time reduction. Your customers will be satisfied, your costs will go down, and your profits will increase. It is my opinion that the future belongs to the flexible. If you are flexible, you have a great future ahead.

CUSTOMER DELIGHT

A great deal has been said lately about customer delight. What is it? What does it mean? Simply put, with set-up time reduction you will be able to meet the requirements of every customer. Customers want four things:

No defects.
On-time delivery.
Correct counts.
Low prices.

If you couple set-up time reduction with advanced quality method, all four criteria will be met. Set-up time reduction must not ignore quality issues. It must be built on quality focus. With set-up time reduction, the delivery will be quick, and you soon will produce in the quantity ordered by the customer with no set-up scrap or rework. Your costs will go down. If the customer needs lower prices, you can pass some of the savings along, while rewarding yourself with additional profits.

Internal Customer Benefits

As important as the external customer delight is to your company, the same is true of the internal customer. If the product is delivered late, is miscounted, or is missing parts, the organization will disappoint the external customer. Having your employees identify their internal customer and determine their customer's needs will go a long way toward achieving external customer delight. Internal customers need the same four things the external customers need:

No defects.

On-time delivery.

Correct counts.

Low cost.

Set-up time reduction provides the same results internally. Employees need to recognize their internal customers.

FLEXIBLE SCHEDULING

Currently, scheduling departments are stressed with the demand for product and the backlogs at each machine. Scheduling managers are constantly adjusting the orders to meet the changing demands of external customers. Lead time, set-up time, equipment breakdowns, and quality problems contribute as well to the enormous task of deciding which job to run next. Once they get it all set, an executive makes a commitment to a customer that puts them back into adjusting the schedule. Sound familiar? That's because you are not alone; it's common in most manufacturing companies. The alternative is set-up time reduction.

Today your schedulers are asked to do the impossible: meet customer requirements without improvements to the production process. Quick setups eliminate many of the problems plaguing the company—free the capacity, allow for maintenance to perform preventive maintenance, and get the product shipped on time without intervention. Flexibility and quick response are the key; set-up time reduction is the driving force.

SAVINGS

If an initiative doesn't save money, it probably isn't worth doing. There are a lot of nice-to-have initiatives that don't show up on the bottom line. These are the programs that come and go through the life of an organization. Set-up time reduction should not fall into this category. If your team(s) will follow the proven methods and not stop improving, and if the organization reduces the lot size as the achievements are made, you will see a contribution to the bottom line in the form of increased profits.

You need to recognize that the savings may come as a result of many teams working on set-up time reduction. Additional savings are normally the result of a new attitude, a reflection of a focus on the cost drivers and a respect for the input of all employees. All of a sudden, many people in the organization recognize the need to make improvements and not let waste happen. Do not be surprised when the savings begin to mount and profits are greater than expected. Recognize that the savings are the result of the team effort; openly recognize the efforts and keep the savings coming.

PRODUCT INVOLVEMENT

This may not be a result as much as a direction given to set-up time reduction. It may be well worth the effort to focus the initial team(s) on one product line that is critical to the future of the company. If so, get the results in the specific product and do not allow the improvement to be improperly focused in the wrong area. If your customers need improvements in a certain area, focus the teams in that area first. The other is the "nice to have" which they can work on later.

CYCLE-TIME REDUCTION

This is the ultimate result of set-up time reduction and will be achieved if you reduce the lot size as a result of reducing the set-up time. Achieving cycle-time reduction is the benefit that will pay back for years to come. Just don't allow the improvement to stop, but keep pushing for continued results from the set-up time reduction teams; then push for lot size reduction until a lot size of one can be produced economically. The ideal lot size is whatever the current requirements are. Current requirements vary by industry type:

Process industries = one day's usage.

Repetitive industries = daily build plan.

Job shops = actual customer order.

With set-up time reduction, you will be able to produce, ship, invoice, and be paid for product in the shortest time possible.

As you can see, there are many improvements you can expect from set-up time reduction. All of the benefits contribute to the customer satisfaction while reducing costs. Your expectations of set-up time reduction should be high since achieving the initiative may be crucial to overall success. Achieving the initiative and realizing the results are well worth the effort required. Expect success and greatness.

CHAPTER 10

The Team Experience

WE NEED MORE RICARDOS

A few years ago I had the opportunity to meet Ricardo Corral. He operates a sheeter, a machine that cuts rolls of paper into the sheets used by printing press operators in a high-quality printing company. At the time, I didn't realize how much he would begin to impress me. Ricardo speaks both Spanish and English fluently.

When we met, at a training session I conducted, Ricardo took me out to the sheeter and pointed out some of the needed improvements. The roll stand was very loose and needed to be fixed before other improvements could begin. I agreed with him, and made a note that we should get maintenance involved. On my next trip to the plant, Ricardo again asked to meet at the sheeter, showed me how the roll stand had been repaired, and then showed me other things that needed to be fixed. As time went on, the same thing happened. He would show me what needed to be done; then *he* saw to getting it done. Sometime later, the manufacturer of the equipment came to the facility to see a machine that was 30 years old holding tighter tolerances than when it was designed, and with greatly reduced make-ready time. This was accomplished as a result of the work of Ricardo and his team.

Ricardo could be called an empowered employee. He works well with other employees, and gets improvements made. Why do some employees grasp an initiative like set-up time reduction and take it to high levels of accomplishment, and some employees say it can't be done? Probably a number of things, so let's look at some of them.

ATTITUDE/ASSURANCE

In order to work effectively for the good of the organization, the employees need to be assured that the improvements will not jeopardize their future. Job security is a real concern today in business, and if you say there are no guarantees, maybe there should be. If you look at the cost of doing business today, set-up time is a large part. If set-up time is reduced and benefits realized, possibly keeping an extra employee on the payroll may not be expensive at all. If people are laid off as a result of team improvements, the improvements may stop. If people have no work to do as a result of set-up time reduction, you now have someone who can work full-time on further reductions.

An employee like Ricardo doesn't concern himself with assurances. He or she simply examines the initiative, decides to give it a chance, and does it. There are Ricardos in every company. All you have to do is get them focused on set-up time reduction. Your goal should be to create a workforce that makes improvements happen.

THE CULTURE

My experience has been that managers usually say they want change and encourage it, but the employees say they don't experience it. If someone tells you we want to achieve set-up time reduction, trains you, assigns you to a team that meets every week, and then never says another word about it, you might doubt their desire to achieve that goal, regardless of how big a deal they made of it in the beginning. Starting an initiative like this is easy. Starting anything is easy and takes very little effort. Starting is not enough.

You need to create an environment that ensures that the organization will be able to achieve the results. The environment consists of actual and perceived factors that need to be examined and in many cases changed. Let's start with commitment.

COMMITMENT

Commitment starts with an honest evaluation of the work that is required to achieve the benefits. If you have not evaluated the work, then you should question your commitment. If you say we'll do whatever it takes and then put off doing whatever it takes there is no commitment. In fact someone who says "I'll do whatever it takes" may not have taken the time to examine the subject and what it will take to accomplish.

In order for an organization to survive in the future, it must commit to the following:

- On-time delivery to customers.
- Elimination of defects.
- Reduction in set-up time.
- Elimination of waste.
- The importance of people.
- Managing and supervising by sound leadership.
- Team problem solving.

It is perfectly acceptable to have more than one important issue. Being committed to something means that when it is time to do it, it gets done. Too often though, there is only an up-front attention and not a lasting commitment.

Shipments are important, elimination of defects is important, managing and supervising are important, and so is set-up time reduction. In order for set-up time reduction to be a success, the commitment needs to include:

1. Team training.
2. Regular team meetings.
3. Assignment completion.
4. Financing.
5. Progress measurement.

AUTHORITY AND RESPONSIBILITY

In order for your employees to achieve the necessary levels of set-up time reduction, they must have both the authority to make change happen, and the responsibility to make it happen as quickly as possible. The authority comes as a result of willingness to apply this method of set-up time reduction. This authority must be real and if the organization does not give the authority, the results will be slow at best. Someone must give up authority in order for the team to get it.

In most companies, one person or department does not have the authority to reduce set-up time. That authority is dispersed among many departments such as tool design, toolroom, manufacturing engineering, shop supervision, and the set-up employees. Your company must make certain that the efforts of set-up time reduction are supported and in no way hindered by these functional areas.

It is important that the team recognize the need to make improvements and achieve the results as quickly as possible. Without being discouraging, you need to put pressure for results on the team. It is easiest and most reasonable if the person pressuring the team is also a team member. The team must not be satisfied with slow improvements. The company must allow and expect change. The entire organization needs to expect change from all the teams and especially the set-up time reduction teams.

TEAM MEMBERSHIP

Teams make better decisions than individuals. Individuals may come up with one way to reduce set-up time, but it may be high cost or it may create more work in another area. Teams will consider more aspects of implementation. There are many great ideas waiting to be completely developed, improved, and implemented, and a team structure enables that process.

Informal barriers exist between people and their departments. Quality, engineering, manufacturing, accounting, purchasing, production planning, stores, sales, and so on, may all experience informal barriers with one another. First shift and second shift employees may have informal barriers. You need to channel this competitive drive for set-up time reduction. Simply establishing teams in your company will begin to overcome barriers. Yes it will take time, but it took time for the barriers to be put in place. As these teams work year after year and progress through the phases, you will begin to see barriers fall away. When you begin to hear the word "we" instead of "they" in conversations, you know you are on the way to making the quantum leaps that will drive the organization to world class.

Pride of authorship is one of the barriers in some companies. Some employees may be offended when you present set-up time reduction as if they had never thought of it. Certainly there are employees in your organization who have had ideas for improving set-up time. Encourage them and get the input during the training sessions. Recognizing the ideas and reinforcing the value they add will go a long way to eliminating the pride of ownership issue.

You may ask, "should we ask for volunteers or hand-pick the team members?" The answer is yes. Make a presentation on the need for set-up time reduction and then ask for volunteers. If everyone needed for the team volunteers, you are ready to start. If not, you may need to talk with key people to get them on the team. Most employees are willing to work on this initiative once their questions and concerns have been answered. Taking the necessary time to put the right team together will be rewarded when the team reduces set-up time.

Typical membership in set-up time reduction teams is not cut and dried, but you should consider representation from the following areas or departments:

- Engineering (design, industrial or manufacturing).
- Tool crib.
- Tool design.
- Material handling.
- Quality.
- Stores.
- Manufacturing (department supervisor, set-up expert, and an operator).
- Maintenance.

You should be sure to include different levels of the company as well, in order to get the mix that the team will need. Team membership should consist of a senior manager, a middle manager, a supervisor, and technicians for understanding from various organizational levels.

Unfortunately, most teams meet on company time. It is unfortunate because it limits the amount of time the team can spend working on the initiative. If the meetings are on company time, they should be one hour long and every team member should take one hour's worth of assignments to be done between meetings. Meetings should start on time and stop on time so the other employees will know exactly how long the team members will be away from their work area. This will eliminate a lot of the frustration associated with teams. Typically these meetings are scheduled weekly.

DECISION MAKING BASED ON FACT

Opinions are a good starting point but they require facts to support them. Your teams should act on facts and therefore need to collect information in the form of data to determine the facts. Data collection is real work and should never be seen as make work. Data are the raw material of problem solving.

WHEN THINGS SLOW DOWN
OR YOU FEEL LIKE QUITTING

It is not uncommon for the teams to start well, but as time goes on, encounter a stage that makes them feel like quitting. When this happens you should ensure that the team process check is being used by the team. This simple technique makes the team help itself. Here's what it looks like.

At the end of the meeting, every member rates the meeting in each category. Any reasons for ratings that are extremely different from the rest must be verbalized by the member making that rating. This process check helps the team members discuss how they are doing and how to help them-

TABLE 10–1

Team Survey Checklist

Team Process Check									
Low									**High**
1	2	3	4	5	6	7	8	9	10
Participation									
Listening									
Leadership									
Decision quality									
On track									
Fun									

selves. Many teams average these scores for an overall meeting score which is tracked graphically.

Another method you can employ is the team questionnaire. It consists of three questions. If you decide to use this questionnaire, every team should be asked the same questions. If you ask them of only the team that is struggling, that team may feel singled out and rebel.

TABLE 10–2

Team Questionnaire

Team:_____

Date:_____

Not considering any individual, please respond to the following questions:

  1. Are you satisfied with the team's progress?

      ☐ Yes   ☐ No

If the answer is no:

  2. What do you need to do to help the team do better?

  3. What does the team need to do to improve?

If the answer is yes:

    What comments do you have?

This questionnaire should be reviewed with the team and the team should determine what, if anything, it needs to do. A team that is doing well may do even better. A team that is not doing well will probably make its own corrections without the intervention of outsiders.

DISTRACTERS, PROCRASTINATORS AND OTHER INFLUENCES

Team members have personalities and because of that, the team may be distracted. It may set off on another project that takes away from set-up time reduction. Distracters must not be tolerated, the team must remain focused and achieve its goals. Any team that is distracted must be managed by the steering committee to get it back on track. Reading the team mission statement at the start of each meeting can help, as will understanding personality types. Some of the problems teams and companies face come because of a lack of understanding of personality styles. It is important that everyone understand there are different personalities. An understanding of the styles may eliminate many of the problems they face.

Let's look at each style and try to understand what makes them offend the other.

The driver is an impatient person who wants to move toward the goal and doesn't tolerate much discussion or alternative planning. Drivers want to know that the rest of the team will work with them in achieving the goal.

The expressive is a vocal and excitable person who experiences severe ups and downs. Expressives need recognition and inclusion. They periodically threaten to quit the team when they are in one of their down times.

The analytical is slow to tire, always wants to collect more data, and is slow to make a decision. These people love charts and graphs and are always asking more questions even though the team may have already made the decision.

The amiable is very people oriented and seems to be more interested in forming relationships than making a decision. These people are interested in the individual before the team. They want to make friends; other issues are secondary.

Once team members understand the personality styles, they realize why members respond differently than might be expected. Personal styles are usually the reason a team flounders. Understanding the styles allows the different personalities to see why they respond the way they do, and to become more tolerant of the other style.

Remember, no one personality style is better than the other, and that may be the reason teams work so well; individual styles make the team move forward at times, collect more data at times, get excited at times, keep everything in perspective at other times, and recognize the feelings of everyone.

MANAGING AND SUPERVISING WITH TEAMS

As teams develop, some of the decisions that formerly were made by management and supervision are shifted to the team. While this may sound positive, it can be disarming to those people who feel they are losing control. This can result in managers who resist the team problem solving and even try to subvert the teams' ideas. Your company will need to be aware of this and work with those people. If they can't give control to the teams, the problems teams are set to solve may never be solved. You want teams that attack problems, implement changes, and are not hindered. At the same time, you need managers and supervisors that get those improvements which result in the attainment of strategic plans. Both are important and must not be overlooked.

Your set-up time reduction team has responsibilties that were once the responsibility of management and supervision. Anyone who sees an opportunity to improve should bring this to the team. Ideas must be respected and acted upon for the good of the organization. The team must take responsibility for getting set-up time reduction for the good of the organization. This takes nothing from the manager or supervisor who is responsible for the day-to-day operation of the department, except that the manager or supervisor now has people helping him to make the area more productive. By properly supporting the team efforts, the manager or supervisor can get more accomplished.

Still overcoming the fact that the team will make decisions and implement changes is difficult for some to accept and may hamper the team's progress. The organization needs to support both while ensuring that the team accomplishes its goals. Set-up time reduction requires support that is true and sincere. Lip service support will distract and impede effort.

WHERE'S THE BAND?

Without outward shows of encouragement to the team, it may cease to thrive. Many companies have a day a month set aside to recognizing the improvements implemented in set-up time reduction. If monthly is too frequent, then once a quarter should suffice to let the teams have that sense of making contributions that are recognized by others.

There are two key ingredients to providing direction:

1. Measurement(s)
2. Results orientation

Measurements are kept by the team and allow it to determine its progress. These measurements should be posted for everyone to see so they too can monitor the team's progress.

The results orientation should be adopted by every employee. Without results the effort will not succeed. Everyone in the organization should recognize the need for results and should put a great deal of pressure on themselves to help achieve those results.

THIS IS MY TEAM, I HAVE OWNERSHIP
TO ITS PERFORMANCE!

Ownership is where you want your teams to get to. Once the team members take ownership of their performance, they will maintain the effort without close monitoring. There are none of the personal attacks that flare in the process. At this point, the team is attacking the problems, not each other. It is a pleasure to observe a team at this stage, because they simply do the job without a lot of fanfare.

At this stage, the team members do what is needed for the good of everyone, not just themselves or their work area. Team members will volunteer for assignments that earlier they would have not taken. Team members will offer to help other team members who are overloaded with assignments.

Another aspect of team maturity is that no one acts on opinion. The team looks for the facts. The team may also go out to the work area more often to make sure the decisions are the correct ones before it decides to implement. The team members will talk to other operators and set-up people more frequently and will see that their decisions may need adjustments as other improvements are made. Caution the team not to be overly concerned since this is continuous improvement. Teams should make the best decision at the time, and if down the road a better decision is possible because of other changes, then so be it. If one waits until the absolute best improvement can be made, no improvements may ever be made.

MANAGEMENT: THE BOSS IS KEY

Management has a vital role to perform in the development of set-up time reduction teams. Henry Ford stated that "The hourly worker will do what you ask them to do, the challenge to management is to ask them to do the right things." Supervisors and managers become critical players in the achievement of set-up time reduction. If they encourage the effort, progress will be made; if they display enthusiasm, everyone will become excited and the progress will be even faster. If they never bring up the issue, it may die a quick death.

Your company can achieve all of the important tasks such as making shipments, achieving high quality levels, and reducing set-up time if everyone sees the need and tries.

In the team environment, individuals become less important, and teamwork becomes more important. But this will only happen if management embraces team problem solving and allows it to happen. You may need to calm the fears of the management and supervisory employees that see this new enviroment as a loss of authority. Tell them this is not a loss; these teams will help them meet the requirements of delivering on time at a lower cost in a reduced cycle time.

SATISFACTION IN A JOB WELL DONE

Satisfaction in a job well done is awaiting your employees as they are allowed to make the improvements. Progress may be slow at first, but rest assured that with the proper support, the success will be great. The satisfaction and pride the workforce displays is well worth the effort.

The steps to success are simple:

1. Decide if set-up time reduction would benefit your company.
2. Train the workforce.
3. Support the effort through all management and supervisors.
4. Institute team problem solving using statistical methods.
5. Track your success.
6. Don't ever stop.

Employees like Ricardo make the process of implementing change through teams fun and easy. Such employees are a valuable resource and a role model for others. Not everyone will support the effort at first, but once they realize the benefit and management commitment, they too will become like Ricardo. Find your "Ricardos" and get started.

CHAPTER 11

How to Make It Fail

Many times people ask what makes set-up time reduction fail. This question is disappointing because set-up time reduction shouldn't fail, but most people are asking to learn from others' mistakes. Some may want to see if the subject is for real; if there are no failures, maybe there were no real successes either. This chapter is dedicated to what you could do to make the effort fail or rather, to provide you with a reverse-negative list should you choose to succeed. The most frequently occurring causes of failure are listed first.

DON'T TRAIN THE TEAM

Training is worth the time and effort. A team trained in the principles and techniques of set-up time reduction will excel. A team without the training may flounder and quit. Set-up time reduction is a focused initiative that is easy to learn. Once the subject is grasped, the team can move forward quickly. Every team member should be trained to prevent the team from having to learn the hard way, through trial and error.

WE DON'T HAVE TIME

A few years ago a plant site had this sign:

NO TQM, SET-UP REDUCTION OR CONTINUOUS IMPROVEMENT TEAM MEETINGS THIS WEEK.

WE MUST MAKE SHIPMENTS!

This was a sure sign of failure; that plant is now out of business.

Don't let this happen to you. You are busy, shipments must go out, and it costs money to make improvements, but the future belongs to the productive. Set-up time reduction will make significant improvements in productivity so you can't afford *not* to do set-up time reduction; the marketplace is too competitive. Your competition hopes you don't take the time to make improvements. It is hard to believe that an employee going to a one-hour meeting and doing one hour of assignments will make or break shipments in any one week. If you feel you may have to put a sign up like this, you should further evaluate the benefits and decide whether you are committed to set-up time reduction.

A FEW PEOPLE ON TOO MANY TEAMS

When you first start an initiative like set-up time reduction, it is difficult *not* to involve those employees who motivate others and always seem willing to work for the organizational initiatives. These people have a positive outlook and usually have success. They probably were the first members when you started other efforts like management by objectives, total quality management, continuous improvement, quality circles, employee empowerment,

11–1   You've set aside a room for the teams to meet. Now make sure it is used.

preventive maintenance, supplier certification, process flow improvement, and inventory reduction. They are always involved when something needs to be improved. Self-starters, they don't listen to criticism. They are positive in their outlook and pleasant to be around. You wish you could clone them. These people often wind up in a leadership role as role models and even trainers, but be careful about over-loading them with too many meetings.

ALLOW MANAGEMENT TO SKIP MEETINGS

Since there are many pressures for management's time, you must decide what is important. If you want to achieve set-up time reduction, the team meeting must be a priority. Decide how to do the other things, implement time management, ask other managers for help, but do not skip the team meetings. If it is not important to you, it will not be important to others in the company.

There are only two times when someone should miss a team meeting:

- When he or she is sick.
- When he or she is on vacation.

Otherwise all team members should be at the team meeting. Travel, other meetings, and daily work must be scheduled around this one hour each week.

DON'T LET THE TEAM MAKE DECISIONS

If your teams are structured properly to include senior management, middle management, supervisors, and the technical experts, it is easy to give responsibility, authority, and money and simply ask it be used wisely. These people will not make dumb decisions. They will get frustrated though if they have to get permission every time they are ready to implement a change.

If the financial impact of the change exceeds the level of approval the team members have, then they must justify the expenditure and submit the proper forms for that approval. Let them make decisions and then hold them accountable.

DON'T TAKE ASSIGNMENTS

If anyone on the team is allowed to refuse assignments, you will find that soon few if any team members will take assignments. Shortly after that, people will want to quit the team and future improvement is virtually impossible. The rule to follow is that every team member will take one hour of assignments from every team meeting unless he or she will be on vacation during the following week.

DON'T GET ASSIGNMENTS DONE

Once the assignments are taken, they must get done. It will become very frustrating to the team if someone constantly fails to complete their assignment. Sometimes an assignment estimated to take one hour takes more time. When that happens, spend an hour on the assignment, report back at the next team meeting, and continue to work on that assignment one hour a week until it is complete.

If you can spend more than one hour on the assignment, you certainly should. Here is the point: One-hour assignments are the minimum. Most employees have more than an hour a week to work on set-up time reduction. When machines are down for one reason or another, that is a great time to work on set-up time reduction. Managers and supervisors have discretionary time—this also is a great time to work on set-up time reduction. Just remember to work on what the team has decided, and not on just what *you* choose to work on.

DON'T DISCUSS IT

It is important for management to discuss frequently the status of set-up time reduction. Likewise it is important that every team member discuss what the team is doing with other employees. Encourage the continuous open discussion of the initiative. The feedback received during these discussions may be helpful to the team as it moves forward. Discussions with other employees also prepare those employees for the changes they will experience and allows them to have input to that change.

Supervisors must understand the need employees have to discuss changes that the team is considering. If two employees are talking about set-up time reduction, it may be best to let them continue rather than criticizing them for wasting time and exhorting them to get back to work. Likewise, employees need to work on production when it is needed, and not work on set-up time reduction all day long. Reasonableness is the key

CRITICIZE INSTEAD OF ENCOURAGING

It is easy to be critical and some people do it without even realizing it. Watch for the verbal and nonverbal feedback and make sure you are not criticizing. I once observed a senior manager questioning a team. That questioning was taken as critical and the verbal response of team members made that interpretation clear. The senior manager was in fact trying to get the team to take credit for a good decision that was difficult to arrive at. The senior manager continued to press for answers to his questions and to this day the team does not respect or listen to this manager. If you need to ask questions for clarification, tell the team what you are doing. The manager who was questioning the team should have prefaced his remarks by saying he didn't feel they were getting all the credit they deserved. Another senior manager at a company made the comment that "set-up time reduction will help us if we don't go broke doing it." Three team members heard the comment and asked to be taken off the team. They did not want to be involved in an initiative that management didn't take seriously. This manager was frustrated because a machine was idle due to a team meeting and he had just received a call from a customer who was upset about a late delivery. If you want set-up time reduction to succeed, separate your daily frustration from the need to implement change.

Tell the team what your assumptions are and make every effort to encourage the other team members. Ask others for help if your questions always seem to be taken as a criticism. In most companies, teamwork is fragile and must be nurtured in order to overcome that fragility.

WORK OUTSIDE YOUR CURRENT SYSTEM FOR APPROVALS

Many companies put the steering committee in the position of approving the team's recommendation. This means the steering committee has to agree with the team. This means the teams have to convince the steering committee of the validity of their ideas. This process is slow at best, and fails in most cases. The job of the steering committee was presented earlier in this book. In order for them to steer the efforts, it is important that your current systems for approvals be followed by the team. Adding another level of approval is not recommended.

You should ensure that every team knows how to get approvals and then expect it to meet the requirements for obtaining them. If you have staffed your teams with the levels of the company involved, this is a simple task. Supervisors can usually approve certain changes, middle managers approve other changes, and senior management still others. These managers will also know what forms are needed as well as the company criteria for approval. If the team's solution can not be approved for any reason, the people on the team should recognize that and either develop alternate solutions or wait until their solution meets the criteria.

DON'T PROMOTE SET-UP TIME REDUCTION ONCE IT IS STARTED

Most companies start an initiative like set-up time reduction with a lot of hype and encouragement. What happens after that is important. If that is the last time top management speaks of the benefit and no feedback is offered, the initiative is sure to fail.

Everyone needs to talk frequently about the initiative— its successes, its failures and the continued need to achieve are important to communicate. If set-up time reduction is worth doing in your plant, senior management should commit to talk about it daily. As conversations happen with employees, they will catch on and begin to bring the subject up themselves in their discussions. If you think set-up time reduction is necessary, then you should say it a lot.

DON'T MEASURE PROGRESS

Some teams find it difficult to develop measurements that track their progress. You should never tolerate an improvement team that cannot measure its improvement. Without measurements, teams either become frustrated or may fail completely. With measurements, teams can track their own progress and know how well they are doing. Measurements allow teams to evaluate the effectiveness of their work, and the company can watch the progress as it is happening. Insist on measurements and look at them frequently. They will "keep you posted!"

MAKE REPORTING DIFFICULT

Some companies seem to want to make the process of reporting progress as difficult as possible. Sometimes this is done by not providing a proper reporting process. Have the teams make monthly progress reports to the steering committee, limited to five minutes, during which the team presents its mission, measurement(s), and decisions since the last report. Reading team meeting minutes to the steering committee is not necessary, but progress reporting is.

DON'T FOLLOW UP OR MOTIVATE

This is very similar to not promoting set-up time reduction once it starts. Taking the time to follow up on the training, team meetings, and individuals involved will ensure that everyone recognizes the initiative's importance and your commitment. Consistent follow-up by interested individuals will do much to ensure that the team(s) are motivated to do well.

Sustained effort is better than a quick start any day. Continued motivation is necessary for that sustained effort. Keeping the vision of the future out front helps motivate. Once your teams start, encourage and challenge them to excel. Some companies identify champions whose job it is to follow up with the team and remove any barriers that might slow the effort. If you are the champion, doing your job will help motivate the team.

MAKE THE FOCUS INDEFINITE THEN QUIT

Stress the need for a results orientation, meaning the set-up time reduction team and steering committee should feel pressure for results. You will want the team to have a definite focus in a particular area. That may be a machine that is the bottleneck to making deliveries on time, a cell that is critical to the future of your factory, or simply a machine that is critical to success. Ask the team to focus on set-up time reduction, and ensure that they remain focused until the job is well done.

If the focus is indefinite in its purpose, the effort will stop. Make the focus specific and press for results in as short a time period as possible. Once accomplished, either change the focus of the team to another machine or set a new goal for the current machine.

DO IT FOR A WHILE UNTIL IT JUST FADES AWAY

Some companies start set-up time reduction with much fanfare, teams meet for a while, and then the initiative begins to fade away. To guard against this, have the team report to the steering committee and keep the excitement in the initiative with continued motivation and pressure for results. There is no good reason for set-up time reduction to stop. Management has a responsibility to become involved and to keep the initiative alive.

HAVE "TEAM OF THE MONTH"

The problem with competition is that there are losers. In set-up time reduction, there are no losers. The entire company benefits from the efforts of these teams. Unless you rotate and everyone gets to be part of the team of the month, do not recognize one team over another. Your competitors are the ones you want to beat. No one within your company, involved in its improvement, should be seen as a loser.

DON'T GO OUT TO THE SHOP TO SEE RESULTS

Too often a team works very hard, achieves results, but no one notices. Ten-minute trips into the area where a team is working is time well spent. Not only will you see first-hand their accomplishments, but they will be encouraged by your interest. No one likes to do work that is never noticed, and it is difficult to visualize an improvement that you have never seen. A team that never gets encouragement is a team that will disband. A trip to the shop is very encouraging.

DON'T RECOGNIZE ACCOMPLISHMENT

It is not uncommon for a person to be vocal to others about the accomplishments of a set-up reduction team, but if the team never hears the recognition, the efforts could stop. Don't let this happen in your company. When you recognize accomplishments, tell the team and be specific in that recognition.

DON'T HOLD THE GAINS

In order to support the continual improvement required to be a world-class company, the improvements implemented must be kept. Many companies allow teams to make improvements, but if those improvements are not held, the team will begin to question its existence. It is probably more difficult to hold the gains than it is to develop and implement them in the first place. Examine your systems and procedure enforcement, and make the necessary changes to ensure that the gains achieved by a team are never lost.

A little caution is in order here. It is normally the responsibility of the supervisor to enforce a new work procedure. Your teams must make it simple for the supervisors to know if the set-up experts on all shifts are following the new method. Make the procedure visual and easy to follow.

DON'T ENCOURAGE LOW-COST SOLUTIONS

As teams begin, you need to stress the importance of developing low-cost solutions to reduce set-up time. If the team has a solution looking for a problem, it may become disappointed as it tries to get approval for funding. At the start, most companies have limited budgets to achieve set-up time reduction. Encouraging and expecting the team to develop low-cost solutions is prudent and reasonable. As improvements are implemented, additional funding for the more costly improvements may become available. A team should never offer only one high-cost solution for any problem. If you follow the problem-solving method outlined in this book, the team will develop alternative solutions.

SECOND-GUESS TEAM DECISIONS

Sometimes the intention is to gain understanding, but if the team perceives that you are second-guessing their decisions, they no longer feel empowered and will look to you for decisions. If that is what you want, don't empower a team; do it yourself. If you question team decisions, you will limit the speed of improvement. Don't make this mistake. If you need clarification make it perfectly clear to the team that you are not questioning the decision, just making sure of the process they used. If your teams are following the proven method presented in the training, they will have considered low-cost solutions.

DON'T FOLLOW A PROVEN METHOD

Provide your set-up time reduction teams with proven methods and techniques. Help them organize and provide them with the tools to make change happen. You can develop your own method for reducing set-up time, but if you do so, fully develop it before you go about implementing it. Take the time to ensure success in the team(s). If set-up time reduction fails because the method didn't work, you will have a very difficult time getting the initiative started again. Set-up time reduction is simply too important to let it fail.

CONCLUSION

I'm sure I haven't exhausted the possibilities for failure and maybe every one of these by itself won't cause failure, but a conglomeration of them will. I hope this hasn't been too discouraging to you since set-up time reduction is an exciting improvement possibility that is easily achieved. It is worth the effort and won't fail if it is supported properly. Do it and give your employees a real sense of accomplishment and security.

CHAPTER 12

Most Frequently
Asked Questions

In this chapter, we will review the questions that are asked most frequently and the responses that you may give. People have concerns when you start an initiative such as set-up time reduction and you should be sensitive to those concerns.

If we make the setup too easy, won't the company just hire cheaper labor to do our jobs? This is always a possibility, but it is not recommended. In most companies, the ability of the set-up expert determines how quickly jobs are set up. Getting rid of that experience would be a shame, much the same as wasting that experience.

Will set-up time reduction eliminate jobs? If it does, then the improvement will probably stop. No one should be eliminated due to any improvement requiring teamwork. The company should determine if it plans to eliminate jobs as a result of set-up time reduction before the first person is asked to be involved. If the answer is yes, then tell the people up front.

I hope the answer is no one will lose their jobs as a result of set-up time reduction. This will allow the employees to attack the problem and not be concerned with job security.

Determine with management whether a policy of no layoffs due to set-up time improvement can be established. If so, try to get this policy in place. Set-up time reduction will require a lot of work; it will take time and money. If no employees are to be let go, the company must get a return on that investment in the form of reduced cycle time, reduced inventory, and higher productivity.

What's in it for me? The ultimate goal is job security due to working for a world-class manufacturer. World class means being able to compete with the best and win. Today, around the world, companies are working to increase business. In some cases, to do that, the company must take work away from other companies. Those companies that eliminate the waste are going to be the survivors. Much of the time spent in setup today is a waste, including the time spent searching, getting, cleaning, assembling, and so on while the machine is stopped. If you can develop ways to get the old job off and the new job onto the machine faster, then you can begin to produce smaller quantities more quickly, match production with customer demand, and become a very efficient manufacturer.

Business is much like athletics. The team in front doesn't worry nearly as much about the competition. They usually concern themselves with their performance and push themselves to higher levels of performance. Get in front, stay in front, and everything else will take care of itself. Effort properly focused is what you have to look forward to, and if accomplished, the company will survive and provide jobs in the future.

Isn't set-up time reduction just a way to get us to work faster? If it were that easy you would have done that years ago. The goal of set-up time reduction is to make you more productive and the byproduct is the elimination of frustration and waste. Actually, the setups will be faster because of organizational changes, equipment changes, and improvements to the set-up components. The set-up employees should then do more setups. Simply working faster will not get the job done. Becoming more efficient is the key—that is "working smarter."

Most set-up employees in your company today have many ideas for improving their jobs, making the work easier and eliminating the frustration they frequently deal with. Set-up time reduction is a vehicle to get those ideas that are already there, along with new ideas from the entire team, implemented. You will be faster, with less effort.

Just moving elements to external doesn't reduce the time, does it? This is absolutely correct and should be stressed in the team meetings when decisions are made to move an element to external. Moving to external is a great first step. The future steps should be to eliminate or reduce the time to complete those external elements. A team should never move a set-up step to external and then not revisit that element for improvement.

How often should the team meet? A minimum of once a week for one hour. These meetings should be organized with an agenda, begin on time, and end on time. The meeting should be focused on set-up time reduction, and the discussions should be open, frank, and nonoffensive. No meeting should waste employees' time. These team meetings should include reporting of progress, determining next steps, and making assignments.

Why do we need assignments? Without assignments very little work would get done. Team meetings become a tremendous waste of time without assignments. If data collection, information gathering, moving hardware, labeling, organizing, standardizing, developing kitting methods, fastener elimination, and so on are only worked on during a one-hour-a-week meeting, set-up time may never be greatly reduced.

Don't see assignments as a penalty, see them as an opportunity. The changes your team(s) implements will make significant improvements. Assignments allow the work to be spread among all the team members and more can get done every week. As the teams develop, you may consider cutting the meeting to ½ hour and make the assignment time 1½ hours per week. You'll see more work get done and the team will be even more successful.

This amount of time is a minimum. As your employees work on set-up time reduction, they will begin to think about improvements often. Any ideas they have need to be brought to the team and then acted on with assignments. Many of the team members could work more than one hour a week on set-up time reduction. Management and supervision should encourage employees to use any available time to work on this initiative, and they should do likewise. Look at your daily activities and find the time you have available, and then take more assignments at the next team meeting. Don't make the team meetings longer; devote more time to implementing improvements.

What if we can't come up with enough assignments for every team member? Then something is wrong with the team leadership. There is too much work to get done to allow a week to go by without every team member having assignments. There are two simple questions that should be asked by every team member throughout the meeting:

1. What do we need to know?
2. What do we need to do?

From these two questions come all the assignments you will ever need, and you will find that your team members will never lack for meaningful assignments.

What should we do if someone doesn't get an assignment done? First of all, find out why. There will be times that the team members have circumstances out of their control that prevent them from completing an assignment. This should be rare.

Assignments only take one hour and that is a minimal investment. If members are consistently developing excuses for not getting assignments completed, the team should deal with the issue. If the team cannot solve the problem, the facilitator should be asked to work with the member; and if that fails, the senior management should get the situation resolved.

Should management do assignments or delegate them? Everyone should take and do assignments weekly. This includes management. If the managers on the team can have others in their area of control complete some of the work that the team assigns, they can take on those additional assignments provided the team knows that the work will be assigned to other employees. Again, this is in addition to the assignments that they personally will do. Completing your own assignments demonstrates commitment.

How do we know if the set-up time reduction teams are successful? By their measurement graph(s). Every set-up reduction team should monitor its progress toward reducing set-up time in the area in which it is working. With measurements, team members can track their own success and satisfaction.

Who should do the training? The trainers must have good presentation skills. Not just anyone can do this. In many companies the trainers are in the human resources or personnel department, but don't overlook the possibility that training skills can be found in any department. Operators and set-up employees may have excellent abilities in this area.

Training is essential to the ability of a team. I recommend that the trainers also work with the team as it starts up. By doing this, the trainers will be able to answer questions in future training sessions. They will also be able to reinforce the training as the team may require during team meetings.

Can we reduce the training time? The set-up time reduction course is divided into two parts. The first establishes the need for set-up time reduction and is a six-hour class called "Developing the Mind-Set." If your employees don't know why set-up time reduction is important, they need this training. The second course teaches and applies the method of set-up time reduction. This is an eight-hour course during which the team begins to document the setup and brainstorm ways to reduce the time.

Do not reduce the training time. If your employees understand the need for set-up time reduction, they only need the eight-hour course. Any reduction will mean that the team will miss information it needs to accomplish the task. If the course length is a problem, then divide it into multiple sessions and spread it out. Training is extremely important to success, don't take shortcuts.

Do we need a consultant? Not in every case, but many companies do. Consultants assist your employees in understanding and implementation. Many companies have successfully invested in consulting and gained an excellent return on that investment. If a consultant is used, it should be in the area of training, team start-up, facilitator and team leader development, and motivation. This will ensure that the initiative will continue after the consultant has completed their work at your facility. If the consultant constantly tells the team what to do, then you will only be successful as long as the consultant is available.

Will the set-up time reduction team be allowed to make decisions? If the team doesn't make decisions, you are wasting your time. You should not start this initiative unless you are going to allow the teams to make decisions. If the teams are properly structured with set-up experience, operators, middle management, supervisors, and senior management, the teams will make wise decisions. You only need to give them four things:

1. Training in set-up time reduction methods.
2. Time to meet and do assignments weekly.
3. Authority to make improvements.
4. Responsibility to achieve results in set-up time reduction.

Why use the team approach? Teams make better decisions than individuals. You need decisions that get implemented and achieve the desired results. Without teams, you may find too many opinions leading to decisions that are rejected by others in the organization. Teams also involve more people so more gets done each week.

Teams are also an excellent method to break down the barriers that exist in every company. Culture barriers, process barriers, and technical barriers exist and must be overcome. Every team member has experience and creativity that is waiting to be tapped for improvements. Teams provide the vehicle for examination and implementation.

We can't get anything done because the team membership keeps changing! Whatever the reason for team membership to change, downsizing, reduction in staff, lack of commitment, changing initiatives, and so on, it frustrates the teams. The steering committee should evaluate the staffing of the set-up time reduction team(s) and make plans to keep that team in place for an extended period of time (one year as a minimum).

How long should a set-up time reduction team continue to exist? How about forever? Continuous improvement means exactly what it says—improvements that are continuous or never ending. In most companies, there is so much opportunity for set-up time reduction that the teams should never stop improving. The moment you stop improving, your competition has the opportunity to pass you by.

As the team goes into areas and works on different machines, the set-up expert from that machine needs to join the team, and once an employee has learned to work in this team problem-solving environment, I recommend that he is always on a team making improvements.

Who should be team leader? Anyone except management and preferably not supervisors. The reason for this is that managers and supervisors are the decision makers on a daily basis, but in the team it is a team decision-making process. Consider having the operators and technicians as team leaders. Once you have trained the team leaders, they will find leading the team effort is a rewarding experience.

Should the set-up expert in the videotape be on the team? Yes. Many times, while watching a videotape, the team will not be able to determine exactly what the set-up expert is doing or why it is done. The set-up expert can explain. In addition, as the team develops ways to reduce the time, the set-up expert can provide valuable feedback. Remember people resist change imposed by outsiders; make your set-up employees team members.

How do we make _____ (name) be on a team? Many companies recognize the need to have a certain employee on a team and may anticipate difficulty getting that person to volunteer. I'm always curious why someone would not be willing to be on a team, but since I'm answering questions not asking them here's what is recommended.

Have someone knowledgeable about set-up time reduction meet with that person one-on-one. Before the meeting starts, assure the person that they will be allowed equal time to present his or her ideas and give them your full attention. Ask the person to hear you out completely before expressing their concerns or questions. Your presentation should cover the following:

1. Definition of set-up time reduction.
2. The team makeup.
3. Commitment.
4. Benefit.
5. The method to be used.
6. Asking them to just "give it a chance."

If this employee is a set-up person, he may be concerned that you are focusing on him and saying he has done a bad job in the past. Assure him this is not the case. He may ask what changes are we going to make? Tell him you don't know, the team will decide that, and if he is on the team, he will have input to the decision.

Let the employee respond openly and agree with him if he makes valid points. Many times good ideas come out of these sessions and in all my experience, the outcome has been that the employee decides to join the team.

How do we know if our managers are committed? Ask them to be on the team. Then watch and see if they come every week to the team meeting with their assignments completed. In addition, listen to what they talk about daily. If it includes set-up time reduction, I'd say they are committed. It is hard to determine the commitment up front—it takes time. Set-up time reduction will not be done in a few months; therefore the commitment must be just as strong a year later as it was when it started. Respect the word of management, and if they say they are committed, expect them to remain committed. Don't expect everyone to become totally absorbed in set-up time reduction. Expect a reasonable amount of effort to be given regularly.

Can we cancel the team meeting today? Sure, but if you do that three times, it's a habit. Why would you cancel a meeting that is going to reduce costs, provide security, and eliminate frustration? Meetings last one hour, and they should start on time and end on time. These meetings are important if set-up time reduction is to be achieved. If I were an employee and the steering committee asked me to be on a team, showed me the benefits, and then someone said the team meeting was cancelled, I would refuse to let the team meeting be cancelled. Set-up time reduction is a greatly needed future benefit.

How do we remotivate a team? Have a refresher training session reinforcing the need and benefit for set-up time reduction. It is not uncommon for a team to wear down and get tired. At this time, have the team review their accomplishments and list them on a flip chart. Post that flip chart and add to it as the team continues to work. Now, create some excitement about what they are doing by announcing their accomplishments throughout the company. Have a set-up time reduction day or even a picnic, where families are included. Applaud the teams when they make presentations to the steering committee, and do things that make it an enjoyable and rewarding experience. Most of all, senior management should talk frequently to team members about their progress, and encourage their efforts by recognizing the improvements.

Should the steering committee approve expenditures? No. Teams should use the existing company procedures for expenditure approvals. Do not add another approval level for set-up time reduction.

Should the steering committee approve changes? No. This is not the purpose of the steering committee and if it is done, the teams will lose interest and fade away. The teams should be given the responsibility and authority to make set-up time reduction improvements. This means that they are required to develop the improvement, justify any expenditure, and get improvements implemented. The steering committee should not second-guess any improvement the team makes.

What is the steering committee's job? To steer the improvement effort. This means they develop the vision for set-up time reduction; establish the team and its membership; direct the effort by providing training for the teams, leaders, and facilitators; and ensure that the team members are working from the charter they have been given. They also ensure the teams are utilizing the methods presented in the training.

How does the steering committee know if the team is doing well? Monthly reviews should be conducted during which the teams present their progress for the last 30 days. These reviews should last 5 to 10 minutes and the teams should present their measurement(s), problem-solving steps, and decisions. The steering committee should ensure that the team is working on the correct project and that it is using the methods it was trained to use. If the team is not making progress, the steering committee should find out why and solve the problem. If the team is doing well, the steering committee should tell it so, and encourage it.

Why have a steering committee? Any important initiative must have leadership. That leadership provides for training, ensures the vision is attained, and decides when it is time to add more teams. Having this committee gives support for the initiative and provides an opportunity for feedback. The steering committee guides the organization to its goal and keeps it on track.

The steering committee should intervene if the team is not working properly, or strays from its charter. Any obstacles the team cannot eliminate should be dealt with by the steering committee.

Should members of the steering committee be on a team? Good idea! This will provide them with greater insight into the team process. It will also help them to see first-hand the opportunities a team provides and to experience the stages of team development. Members of the steering committee should be careful when its team is making a monthly presentation to let the person presenting do the presentation. When the steering committee is meeting, they are members of the steering committee, not a team presenter. If the steering committee members are not careful, they will overdo the presentation by their team. As the number of teams grows, there will be teams that don't have a steering committee member. That team will be disappointed if they do not receive their due recognition.

Who should be on the steering committee? The steering committee should consist of no more than nine people and every level of the organization should be represented. Senior management, middle management, supervision, group leaders, and the hourly workforce should all be represented.

I recommend that two hourly workers be on the committee. If you have a union, it should be represented as well.

Everyone, when asked for input, will base that input on his or her experience and understanding. It is important that you tap all the experience from each level of the organization. That input is valuable. In the steering committee as with the teams, no one is the boss; everyone is a member and deserves respect and attention when speaking.

What should we do if employees won't allow videotaping? It is probably because they don't understand why videotaping is beneficial or because they feel that the tape will be misused. If there is resistance to video, meet with the people resisting it, listen to their concerns, and decide if they are valid. If so, decide how they might be overcome.

Most employees are against videotaping for two reasons:

1. If they err, they will be reprimanded, disciplined, or negative information will be placed in their personnel file.

2. The tape will be used to train others to do their job.

If either of these two things may happen, you can't blame them for resisting. Do an evaluation on the benefit of the concerns. Neither of them hold as much benefit as set-up time reduction. I recommend that you give your assurance that neither of the two will happen; then make sure that they don't. Employees will help their company provided it is beneficial to both.

How do you get the union to support and participate? The first important step is to ensure that no one violates the contract. Meet with the union representatives and answer the questions they have about set-up time reduction and respond to the issues they bring up. If you are open and recognize the need of the union to ensure that jobs won't be reduced as a result of set-up time reduction, you should have no trouble with them giving set-up time reduction a chance. In truth, set-up time reduction may provide the job security they seek to obtain for their members.

If set-up time reduction team members find opportunity to reduce *run* time, should they work on it? No. Reducing the run time is extremely important, but if the set-up time reduction team begins working in other areas, its efforts will be diluted or it may get totally off track. Whenever the team finds opportunity outside of its charter, it should present that to the steering committee. They will determine who should work on it. Possibly the supervisor from the area would want to get the appropriate people to work on run time reduction. If an opportunity to reduce run time presents itself, make sure someone else develops and implements that improvement, while the set-up time reduction team works within their charter.

Why is it that we've worked so hard and our measurement doesn't show much success? Almost without fail, companies have a difficult time recognizing improvement and understanding the amount of improvement they have achieved. A few minutes reduced multiplied by the number of times you do a setup will result in significant improvement over the course of a year. Most teams measure the average set-up time in the area in which they are working. If the measurement graph begins to show a reduction no matter how slight, the team should try to equate that improvement to an annual financial measure.

The team should know what the monetary value of one minute of set-up time is, and multiply the number of minutes the team has reduced the setup by the number of setups that will be done in 12 months. Multiply that number by the value of one minute of set-up time. This gives the annual financial benefit. Most teams then recognize that the contribution on an annual basis is significant.

On many occasions, companies have been pleasantly surprised to find that at the end of the first year of set-up time reduction, the bottom line shows significant improvement.

Unless you have financial measures for the teams, however, you may not be able to recognize improvements. Many companies have achieved a 50 percent reduction in set-up time the first year. This is significant and will result in profitability gains.

Why do we need to have measurements? If a team has no measurement at all, it will probably become discouraged and stop meeting. You must insist that the team recognize the improvements it has made. Measurements are vital in continuous improvement; they show what the team has accomplished. With measurements you have the possibility of the improvement never stopping. Measurements provide motivation for a self-motivated team.

Why don't our savings show up in profits? Management is responsible for profits; teams are concerned with cost reduction. The management of your company has the opportunity to allocate funds during an accounting period. If your company recognizes the importance of set-up time reduction, it may choose to invest additional funds into the improvement from money saved by the current team. Management may approve expenditures beyond the budgeted amount because of savings by the team. Management may decide to funnel funds into other improvement efforts like statistical process control, preventive maintenance, or supplier certification. You should never see this as negative, because the money was made available owing to your team's improvements. Some companies have invested in new equipment since money was saved through reduced inventory, as a result of set-up time reduction.

It is possible that your company will invest toward the continuation of these very important issues, but I can assure you that if the teams are getting results, costs will go down. Lower costs create jobs and security. In set-up time reduction, don't worry so much about short-term profits. Concern yourself with the need to become a world-class competitor that can set up quickly.

How do we get the rest of the employees to follow the new set-up procedure? Documentation is the first step. The team must provide the proper document that tells the set-up employees how to do the set-up with the improvements. The new method must be written and distributed, and training in the new procedure should be provided. Supervisors should be included in the training, so they too understand the new method. Once that is done, the measurement(s) and communication should tell the team if there is a problem.

Expect problems to surface and when they do, have the team solve them. Follow up with the other set-up employees and get input, answer questions, and generally help them adapt to the new procedure. In many cases, these people may have additional ideas for improvement. Supervisors should ensure that the new procedure is being used. If there are problems, the team should be notified and should then address those problems with solutions.

Why doesn't the company just buy the newest machine that is quicker to set up? Probably because of the cost involved. If you wait until the company purchases all new equipment, you will never get set-up time reduction and the job security that comes with it. Machines are expensive and most companies don't have the resources sitting around to be spent on new machines. Accept the fact that if you reduce set-up time on the existing equipment then maybe the company will have the funds to purchase a new machine. Many companies have proven that set-up time can be reduced no matter how old the machine. The results are well worth the effort and new machines are not necessarily the answer.

Why does it take so long to achieve set-up time reduction? This depends on your perception of time. If you want the setup to be reduced in half by Friday, you are expecting too much. Look at it this way; how long has your company been accepting the current set-up time? 5 years, 10 years, 15 years or longer probably. How long should it take to reduce the set-up time by 50 percent? Let's say it takes 12 months.

That is not very long compared to the length of time setups have been done without improvement. Time becomes relative in that the improvement is normally quicker than the accepted past practice. If it takes a year to make a significant improvement, that was a year well spent.

Rather than being concerned with how long it takes, I recommend that the team commit to making improvements every week. Those weekly improvements will accumulate significantly. Sometimes we criticize people who want to lose weight quickly when it took them years to put the weight on. Set-up time reduction is much the same. Don't worry so much about how long it takes, but worry about when you will start and whether you are making progress weekly. The rest will take care of itself.

What kind of camera should be bought? The camera should be lightweight and portable, with the feature that indicates time on the picture. The preferable time feature is hours, minutes, and seconds. In most cases, this is an add-on item. Also consider the cost of extra batteries and the amount of time each battery lasts when it is fully charged. Having local service support should also be considered in the purchase. Local support may be required to learn to use the camera and for repairs if the camera is damaged.

What is the difference between this and other programs? Set-up time reduction is a focused effort. The result of set-up time reduction will contribute directly to profits. Set-up time reduction is wholly focused on reducing the set-up time in order to produce more frequently, thereby reducing costs.

If you have other problem-solving initiatives established in your company, you should use the same problem-solving tools that are used in the other initiatives, but have the set-up time reduction team focus their efforts solely on set-up time reduction.

What if we have no tooling budget? With or without a budget, your company spends money regularly on tooling.

The set-up reduction team should not concern itself with where the money will come from. When it needs financing for tooling, change parts, or other improvements, it simply fills out and submits the proper paperwork. The problem many employees have is that they *perceive* there is no money available when in fact, there is plenty. Companies are willing to invest if there is a return for that investment. Set-up time reduction certainly returns the investment many times over, year after year.

What if we can't videotape? When you can't videotape, the documentation of the current method of setup becomes more time consuming, but poses no problem. The team members need to observe the current setup and document all the steps taken as if they had a video (this may require that they observe four or five setups). Once it is documented, the team can analyze the setup. Videotaping makes the initial documentation easier, but does not affect any other aspect of the team's efforts.

What if set-up time reduction doesn't work? Then you will be the first company in which that happens. Every company that has tried it has examples of success. The question is, will you stick with it long enough to get the real benefits and the lasting results? These benefits will allow your company to outdistance the competition in productivity and cost reduction.

What if people are against our changes? Early on I said, "people don't resist change; they resist change imposed by outsiders." Suppose you were doing your job and a team came along and told you to do it differently. You too would resist that change and may even go about proving the team wrong. The easiest way to make change happen and overcome the resistance is to involve those people affected by the change by putting them on the team. Talk to them, ask for their ideas, ask them to help with the research, and you will find that the resistance virtually goes away.

Department meetings wherein the team solicits ideas for improvements or needs of the department also provide for acceptance of change. If employees feel involved and a part of the change process, there will be little, if any, resistance. By the way, don't forget to involve the supervision and management. They are just as important to making change happen as the set-up employees and operators.

What does low cost mean? In its simplest form, it means spending a little money and getting a high return. Every team should do a cost benefit analysis when it is planning to spend money. The cost and benefits should be categorized to make it simple, and I recommend that you simply consider three things in both cost and benefit:

1. Labor.
2. Material.
3. Equipment.

When the team considers the cost of the improvement, it should add up all the labor costs, all the material cost, and all the equipment costs. Then it should add up all the annual labor benefits, material benefits, and equipment benefits. Divide the total benefits by the total costs. A good rule of thumb to use is that a result of 20 percent or less is low cost. In addition, there will probably be nonfinancial benefits such as on-time delivery. These should simply be listed, and acknowledged as a benefit that may not be accounted for financially.

How much will set-up time reduction cost? A lot less than the benefits you will receive. Your company should track all costs associated with the implementation including:

- Training.
- Consulting.
- Team meetings.
- Completing assignments.
- Purchases.

- Steering committee meetings.
- Videotaping.

You may even estimate these costs before you start and determine what the results need to be to offset the costs. As improvements to set-up time are implemented, the costs should be compared to the benefits. Don't become overly concerned with the costs, but press for results. The more a team accomplishes in set-up time reduction, the more payback you will receive.

What does empowerment mean? *Employees working in teams, using problem-solving tools to achieve a higher level of productivity.*

The important aspects of empowerment are:

- Teams.
- Problem-solving tools.
- Increased productivity.

What do we do if management won't let us work on the important things? Get involved to change that situation! Many employees perceive that they are not allowed to work on certain things or in certain areas. If you feel these boundaries exist, they will prevent you and your team from making the progress that is desperately needed. An empowered team needs to be focused and use the tools to increase productivity. No manager should restrict a team from helping the organization become more productive. Teams should develop low-cost solutions that increase productivity, and there should be no restrictions on what they work on as long as they work within their charter.

Employees at one company in South Dakota made this statement, "We are only allowed to make the simple decisions like where to hang the brooms and dust pans. The important decisions are made by management." Management assured them that it wanted the team to do whatever was necessary to make set-up time reduction happen. There were no restrictions except in the minds of the employees. Perceived boundaries are just as restrictive as actual ones. Eliminate the perceived boundaries!

Does set-up time reduction ignore safety? Not in any way, shape, or form. If a set-up time reduction team creates safety hazards, it has not made an improvement. Safety is foremost in most companies and that should not change. You must not tolerate any unsafe conditions, and in no way should an improvement create an unsafe condition.

Should we wait until after the layoff to start? If your company is planning a reduction in force at the time it is considering the start-up of set-up time reduction, then the answer is yes. You should make every effort to ensure that no layoffs result from set-up time reduction. If the cause for layoff is set-up time reduction, then something is wrong with the decision. Set-up time reduction should result in smaller lot sizes with the need to do setups more frequently.

What should the set-up time be? It is impossible to say. Don't concern yourself with what it should be, but concern yourself with whether or not your team is reducing the time. The level of achievement is only restricted by the amount of improvement that's been implemented. Companies today have setups that take hours. If that time is reduced, they become more productive. Becoming more productive is what world class is all about.

Is this going to last? That's up to you and your team(s). Your attitude and your opinions will affect the longevity of set-up time reduction. If you believe that it will not last, it probably won't. If you believe it will last and are willing to make that a reality, it probably will. Improvement is fragile in most companies, where programs start and stop without taking the improvement to the higher level.

You need to decide if you are going to make it last. Everyone involved should be responsible to make it a success and to ensure that it is a successful program.

Is management behind this program? They should be, but if you don't know, ask them and then hold them accountable for what they tell you. My biggest concern is not with senior management, but with middle management and supervision. These two levels sometimes confuse the daily tasks of getting orders shipped with continuous improvement. If anyone prevents employees from attending their team meetings or otherwise getting their assignments completed, they are not behind the program and could cause it to fail.

It is very important that everyone from every level understand the importance of spending two hours a week on improvement. If you do not improve, you may be one of those companies that ceases to exist.

If _____ (name) doesn't agree, can we still do it? It depends on the reason for disagreement. Many teams keep from making mistakes by listening to their team members and other knowledgeable employees. If the person has a valid reason for disagreeing with the team decision, you will do well to listen, and in light of that understanding, determine if a better decision is in order. If the person simply doesn't want to change or has no valid reason to disagree, the team should move forward.

The more involvement you have during the problem-solving phase, the more acceptance you will gain during the implementation phase. Involve those people who seem negative, and the possibility of overcoming the resistance is greatly increased.

How long will it take? Don't look at set-up time reduction as something that is accomplished, but rather a process that can always be improved upon. There is no finish line. Move ahead of the competition and, once there, stay in front by continuous improvement. Don't concern yourself with when you will finish—how soon you start is the real concern.

What if we don't have time to do this? Then you are not alone. No company has the time to do set-up time reduction; they *take* the time. Set-up time reduction is an investment and no investment is made without difficulty. There will always be concerns about the resources required, and there will never be a perfect time to start. So just start and limit the amount of human resource allocation to two hours per week. As improvements are realized, you will be pleased that you made the decision to start.

Does set-up time reduction apply to both a large and a small company? Yes, the size of the company doesn't affect the need to reduce set-up time. Productivity improvements in all companies will have a great impact on their futures. Small companies may be limited in the number of teams, but for both small and large companies, set-up time reduction is very important and has application to all manufacturers.

Does set-up time reduction apply to a job shop? Yes. In a job shop the team should make improvements to the machine, not to a specific part that is run on the machine. The focus of the team is to reduce set-up time to increase available run time on that machine. It is important in a job shop that the engineering support be involved in the set-up efforts to ensure future jobs can be changed quickly. Decisions about tooling and fixtures should be made with set-up time reduction in mind.

Does set-up time reduction apply to a food processing company? All of the principles contained in this book apply to food processing although those companies normally call setups "changeovers." Whether it is a flavor change or a complete product changeover, set-up time reduction applies. Wash downs should be moved to external as much as possible through duplication of the change parts that require cleaning.

Does set-up time reduction apply to a printing company? Printing companies refer to setup as "makeready" and that is the only difference. As in food processing, duplication of the parts that require cleaning will allow you to move those tasks to external. All the principles included in this book apply to printing companies.

Can we do set-up time reduction with poorly maintained equipment? Yes, but the company should establish a preventive maintenance program since any improvement in set-up time would mean even more if the equipment is well maintained. Do not ignore the lack of maintenance. Provide the labor, material, and equipment required to keep your assets in top repair.

Who should approve the changes that the team wants to implement? Use your normal approval process. If the supervisor approves certain changes, then the team must obtain that approval. If a middle manager has the approval authority, then the team must obtain the approval from that manager. Don't institute new approval channels. The team should use the current channels for financial approvals also.

How much does set-up time reduction cost? Set-up time reduction should be low cost. This means that the cost to achieve is less than 20 percent of the benefit. Your teams should recognize the need to develop and implement low-cost solutions that provide a payback quickly. Do not allow the teams to identity solutions first. Start with a problem, follow a problem-solving method, and identify the low-cost solutions.

Why should we make more profits for the company? Are they going to share them with us? The impression here is that the employee sees the company as some sort of enemy and not a provider of jobs. Without profits, the company won't exist, so the alternative to helping provide profits is losing jobs. Set-up time reduction will help companies become more profitable. Stop looking at what the company earns in profits as a negative situation. Helping the company become more profitable is an excellent form of job security that is worth the effort. Future generations need profitable companies to provide their jobs as well.

What is a facilitator? The facilitator is the person who oversees the implementation. Primarily the job is to lead the team through start-up and ensure that the teams use the method; once the team chooses a team leader, the facilitator trains the team leader. In most companies the facilitator tracks the overall progress of all the teams, and recommends to the steering committee when it is appropriate to start new teams.

What is the job of the team leader? The job of the team leader is to simply lead the team process which includes:

- Starting the meeting on time.
- Ending the meeting on time.
- Developing the agenda.
- Staying on the agenda during the meeting.
- Resolving any team problems.
- Seeking help for the team when needed.
- Ensuring everyone takes and completes assignments.

What is a champion? The champion is normally chosen by the team from the management level and is chartered to remove any obstacles that prevent the team from doing its job. This includes providing resources, support, and encouragement. Champions don't attend every team meeting, but they are always available when the team needs them.

There may be other questions, which have not been addressed here, but to my mind, the following may be the real question:

"Will you give set-up time reduction a chance?" If it succeeds, you will be part of that success and during the process all questions will be answered and all concerns will be overcome. The future belongs to the productive, and I sincerely believe set-up time reduction will be key to your success. Welcome to the future.

INDEX